North American Mixing Forum 2017

Core Programming Area at the 2017 AIChE Annual Meeting

Minneapolis, Minnesota, USA
29 October – 3 November 2017

ISBN: 978-1-5108-5795-7

Printed from e-media with permission by:

Curran Associates, Inc.
57 Morehouse Lane
Red Hook, NY 12571

Some format issues inherent in the e-media version may also appear in this print version.

Printed by Curran Associates, Inc. (2018)

For permission requests, please contact AIChE
at the address below.

AIChE
120 Wall Street, FL 23
New York, NY 10005-4020

Phone: (800) 242-4363
Fax: (203) 775-5177

www.aiche.org

Additional copies of this publication are available from:

Curran Associates, Inc.
57 Morehouse Lane
Red Hook, NY 12571 USA
Phone: 845-758-0400
Fax: 845-758-2633
Email: curran@proceedings.com
Web: www.proceedings.com

TABLE OF CONTENTS

Some abstracts in this book refer to images that were not made available by the publisher

500914 3D Printing through Chaotic Mixing

Abstract:

Chaotic mixing has the ability to create complex and predictable structures. Here, we use this ability to develop a new 3D printing strategy that we refer to as 3D chaotic printing. A drop of â€œinkâ€ (i.e., a drop of a miscible liquid, fluorescent beads, or cells) is injected into a viscous and Newtonian cross-linkable liquid, and a chaotic flow is applied to generate a complex structure in just a few flow applications. This structure, which arises as a result of the rapid alignment of the injected material to the flow intrinsic skeleton (or flow manifold), is preserved with high fidelity and reproducibility by crosslinking or curing the material. This exponentially fast creation of structure (hundreds of linear meters in a few flow applications), and the accompanying rapid decrease in the length scales of the microstructure (while still preserving high resolution), is not currently achievable by any other 3D printing technique.

In particular, we explore the use of a mini-journal bearing flow and a mini-blinking vortex to implement this novel printing technique. Results from experiments and CFD simulations are used to demonstrate the robustness and reproducibility of 3D chaotic printing.

We envision diverse relevant applications for this technology, including the alignment of fibers or nanoparticles and the development of highly complex multi-lamellar and multi-component tissue structures for biomedical applications.

495735 Improved Mixing in a Pressure Driven Straight Microchannel Using Electroosmosis

Abstract:

We propose a design for improving mixing in the flow of electrolytic solutions in microchannels, through a combination of electroosmosis and pressure driven flow. We show that a secondary flow is induced on application of a transverse electric field. The resulting flow field consists of a pair of helical vortices that transport fluid elements across the channel cross-section. The electroosmotic vortices described here are stronger in smaller channels, due to the increased influence of the electric double layer. We characterize the strength of the circulatory flow as a function of the applied electric field and fluid properties. We then apply this principal to design a straight channel chaotic micro-mixer. Here, we show how chaotic advection may be induced in straight rectangular channels, by applying an electric field whose direction varies periodically along the length. This electroosmotic chaotic mixer has the advantage of efficient and controlled mixing even at low flow rates and small channel dimensions. Moreover, it eliminates the two major drawbacks of active mixing by improving mixing *without* increasing the pressure drop or requiring moving parts. Extent of mixing is characterized by using Poincare maps, Shannon entropy and species transport equation.

502850 Study of Residence Time Distribution in a Taylor-Couette Reactor

Abstract:

Study of residence time distribution in a Taylor-Couette reactor

Haoyu Wang, Anh Phan, Vladimir Zivkovic, Kamelia Boodhoo*

Chemical Engineering and Advanced Materials, Merz Court, Newcastle University NE1 7RU Newcastle upon Tyne, United Kingdom.

**Haoyu.Wang@newcastle.ac.uk*

The flow between the gap of two differentially rotating concentric cylinders is called Taylor-Couette (TC) flow. TC flow exhibits different flow regimes (Taylor vortex flow, Wavy vortex flow, Modulated wavy vortex flow, Turbulent Taylor vortex flow) as the rotational speed of the cylinder is increased (Figure 1). Residence time distribution (RTD) is an important parameter in reactor design which characterizes the flow and mixing behavior of reaction components, thereby influencing the selectivity in chemical reactions. Very few studies have been reported on RTD characterization of TC flow, focusing primarily on limited axial flow rates and rotational speed ranges where the Taylor vortex flow regime is achieved[1,2]. The present work aims to evaluate RTD across the whole spectrum of possible flow regimes in TC flow in order to provide a more comprehensive understanding of the mixing behavior inside a TC reactor.

Figure 1. Example flow regimes of Taylor-Couette flow: (a) Taylor vortex flow (b) Wavy vortex flow (c) Modulated wavy vortex flow (d) Turbulent Taylor vortex flow [3]

Figure 2. Schematic of TC reactor.

The experimental facility consists of two horizontally placed concentric cylinders of 300 mm in length and a gap of 10 mm between the cylinders (Figure 2). The outer Perspex cylinder (50 mm in diameter) is fixed and the inner cylinder (30 mm diameter and made of 316 SS) rotates. The rotating speed of the inner cylinder ranges from 0 to 1500 rpm. The working fluid, a mixture of water and glycerol, is injected into one end of the device at different flow rates of up to 500 ml/min. Pulse injection of the tracer KCl solution at the inlet is achieved by a syringe pump. An E61M014 conductivity probe of 4 mm in diameter and 103 mm in length that connects to a CDM210 conductivity meter is placed at the outlet of the reactor for the measurement of conductivity of the working fluid. The conductivity can be used to obtain the residence time distribution. Initial experiments using pure de-ionized water indicate that the RTD profiles vary between different flow regimes at the same axial flow rate highlighting the mixing behavior in a TC reactor can alter from near-plug flow reactor to mixed flow behavior as rotational speed increases. Further RTD results will be presented and discussed in the context of the flow regime generated in the TC device.

Acknowledgements

The authors would like to thank the European Union's Horizon 2020 research and innovation program for financial supports (grant agreement No 680565).

References

1. Kataoka, K., Doi, H., Kongo, T. and Futagawa, M. (1975) 'Ideal Plug-Flow Properties of Taylor Vortex Flow', *Journal of Chemical Engineering of Japan*, 8(6), 472-476.

2. Pudjiono, P.I., Tavare, N.S., Garside, J. and Nigam, K.D.P. (1992) 'Residence time distribution from a continuous Couette flow device', *The Chemical Engineering Journal*, 48(2), 101-110.

3. Fenstermacher, P.R., Swinney, H.L., Benson, S.V. and Gollub, J.P. (1979) 'Bifurcations to periodic, quasiperiodi, and chaotic regimes in rotating and convecting fluids', *Annals of the New York Academy of Sciences*, 316(1), 652-666.

497959 Experimental Investigation of a New Taylor-Couette Cell Design with Radial Fluid Injection for Controlled Mixing Applications

Abstract:

Flow between rotating concentric cylinders, called Taylor-Couette (TC) flows, are ideal for study of mixing and composite material assembly due to their wide variation in hydrodynamic flows. TC cells have been shown to improve many different processes from cell growth, food processing, extraction, to composite material synthesis because of the controlled hydrodynamics within the annulus. However, traditional TC cell design limits the ability to combine and mix solutions while the cell is operating, due to geometric confinement and complexity of injection when both cylinders are rotating. Here, we present a new Taylor-Couette cell design that allows for radial injection of fluids from the inner cylinder surface into the annulus while both cylinders are rotating. This narrow gap TC cell has 16 (4 sets of 4) injection ports distributed along the inner cylinder. With this geometry, we discuss how radial injection rate and total injected mass during cylinder rotation modifies the vortex structure for two different flow types (stationary and wavy vortex flows). We explore the stability of laminar and turbulent versions of the Taylor vortex and Wavy vortex. Additionally, we explore the effect of increasing Reynolds Number (Re) within the same vortex structure. Finally, the effect vortex structure and stability, turbulence, and Re on the mass transport of the different vortex structures is discussed, where higher order states were found to be more stable to mass injection than lower order states. Optimal conditions for mass transfer can then be identified for future applications.

493574 Experimental Investigation of PVA – Borax Gel Reaction Flow and Mixing in a Non–Element Mixer

Abstract:

Flow characteristics and pressure variation with PVA–borax gel reaction in a Non-element mixer were experimentally investigated. The Non-element mixer consisted of a main flow pipe (inner diameter 10 mm) and three branch flow pipes (inner diameter 4 mm) which are respectively named B1, B2, and B3 from upstream side. The working fluids are 10 mass% polyvinyl alcohol (PVA) solution and 3 mass% borax solution. When these two fluids are mixed, a gel is formed under crosslinking reaction. PVA solution flows in main flow pipe with a steady state condition and borax solution which is colored by blue dye to visualize flow pattern is injected from B2. At first, the experiment is conducted to clarify the relation of flow pattern and pressure variation with gel reaction in the Non-element mixer. Reynolds number of main flow (*Re*) is varied from 0.5 to 30.0 and the injection velocity from B2 (v_{inB2}) is varied from 7.5 to 60 mm/s. Injection time is varied from 4 to 32 s to supply same volume of borax solution to the main flow pipe. Gage pressure variation at 100 mm upstream from B1 is measured by a pressure sensor. The conclusions are as follows.

Three distinct flow patterns with gel reaction are described as, 1. Capsule flow, 2. Capsule + stretched flow and 3. Capsule + stretched + fingering flow. In the capsule flow, the injected borax forms a block namely "capsule" and it flows to downstream direction. The pressure increases when the capsule is formed. However it recovers to the initial pressure when the capsule is discharged. The capsule is produced by gel sheet which is generated the interface between the injected borax and PVA solution. As a result, the injected borax is almost completely separated from PVA region. In the capsule + stretched flow, the capsule is formed in the tip of injected borax and stretched by shear stress of main flow. Therefore gel is generated and adhered on upper side of flow path. In the capsule + stretched + fingering flow, the injected borax forms a capsule and is stretched. Additionally, fine finger shape product is observed ahead of the capsule. Interface between borax and PVA is unstably spread due to viscosity difference of two fluids. This phenomena is similar to viscous fingering. After that, gel is adhered on the main flow pipe. The pressure cannot recover to the initial pressure because formed gel is adhered to the wall of main flow pipe. The flow patterns can be determined by the Input borax fraction and the Mixture velocity. The Input borax fraction is the flow ratio of borax solution and the total flow rate of PVA and borax solutions. The Mixture velocity is a total velocity of main and branch flows. When the Input borax fraction > 0.4 and the Mixture velocity > 40 mm/s, the capsule flow is stably formed.

In order to increase contact surface area between borax and PVA solution to enhance mixing and the gel reaction under the capsule flow condition, we have conducted the experiment using B3. The borax injection velocity from B2 (v_{inB2}) is 30 mm/s, *Re* is less than unity which satisfies capsule flow condition. The injection time of B2 is 8 s. After the injection from B2, PVA solution is continuously supplied to the main flow pipe from B3. The injection velocity from B3 (v_{inB3}) is varied from 7.5 to 240 mm/s. When v_{inB3} is increased such as 240 mm/s, the discharged product forms a sphere which the blue borax solution is covered with thin gel sheet and the pressure decreases to negative pressure when the spherical product is discharged from the main flow pipe. The shape of discharged product and pressure variation can be qualified by first Damköhler number which is the ratio of characteristics of flow time and chemical reaction time.

490210 Industrial Photo-Chemistry in Agitated Reactors

Abstract:

Agitated photochemical reactors have found their way into the industrial chemistry to produce special polymers and rubbers, additives and intermediates, surface active substances etc. The design of such reactors may deviate from the standard. Their profitable, safe and reliable operation is based on some design principles which are explained below.

Chemical reactions need activation energy to transform reactants into products. Catalysts reduce this energetic barrier, but the catalyst is an additional component which is incorporated into the reaction system. It is often expensive and has to be handled before and removed afterwards from the reaction products, where it bears operational risks by its toxic and sometimes pyrophoric character.

Light is an elegant alternative source of the activation energy. Photochemical reactions can be initiated by irradiation with wave lengths of 200 to 700 nanometers, that is from the ultraviolet to the visible range. The photons get absorbed by the molecules, which increases their energy level, thus initiating the reaction. This allows lower reaction temperatures, milder conditions for the conversion of temperature sensitive molecules and finally less by-products.

Typical examples of industrial photo reactions are chlorinations, sulfochlorinations, sulfoxidations or nitrosylations. Loop reactors have been used in the past as a solution for these reactions with their primary liquid phase. However, if gases or solids are additionally introduced, the loop reactor has massive drawbacks in operation, which restricts its productivity.

The new concept of agitated reactors is ideal for these two to three phase reactions. It combines high productivity and flexibility with a reliable operation of the submerged UV sources. Reactor sizes up to 50 m³ or beyond can be realized with the proven design.

In the lab scale, photo reactors are made of glass: the irradiation is simplified due to the small dimensions. The full volume of the reactor can be covered by the light. Production reactors are made of steel, alloys or titanium, and the UV-sources are mercury or LED lamps submerged into the reaction mass inside a quartz glass tube. The glass tubes pass through nozzles in the tank head, where they are connected to the power supply and the cooling medium. They need a hermetic sealing and monitoring to ensure, that in the unexpected case of a glass breakage this is immediately detected. Special care has to be taken during design and installation of the tubes so as to not creating local stress, allowing for thermal expansion and preventing it from vibration excitation, which can develop due to the intense flow produced by agitation. As the power of the UV-lamps ranges from 5 to 60 kW each, the lamps themselves need to be cooled individually inside the quartz tubes, otherwise high surface temperatures could damage the products. Industrial reactors can be equipped with a number of 4 to 20 of such quartz tubes reaching from the tank head into the liquid.

The penetration of the light into the reaction mass is only a few tens of millimeters. Thus, the primary task of the agitator is to provide high pumping rates and thereby a permanent renewal of the reactants in the high irradiation zone

around the tubes. If one of the reactants is a gas as Cl_2, O_2 or SO_2, it must be dispersed into the liquid creating a maximum surface for the gas-liquid mass transfer and dissolution of the gas. Special designs for the primary dispersion of the feed gas and the recirculation of unconverted gas from the head space through the shaft of the agitator, into the slurry, avoids a mass transfer limitation. The homogeneous suspension of solids as, e.g. polymer particles, and the removal of the reaction heat have to be considered as well in the design of the agitator and tank.

The examples above illustrate that often toxic and corrosive reaction components are involved in the industrial photo chemistry. The sealing of the rotating agitator shaft towards the atmosphere is a key element for a safe and reliable operation. Even if temperatures and pressures of photo reactions are low, only double acting mechanical seals or hermetic sealed magnetic drives are applied. The inherent safety of the seal system is based on the monitoring and the automatic refill of seal liquid to compensate for the defined leakage through the seal surfaces.

The presentation provides a concise insight into the potential of industrial photo reactors and proven design principles regarding the irradiation source, agitation and safety aspects.

486416 Characterizing Mixing Processes Using Computational Fluid Dynamics and z-Transform

Abstract:

Computational fluid dynamics (CFD) has now become widely used in studying the performance of mixing unit operations. It complements experimental studies by making accessible to the investigator quantitative measurements and insights that are difficult or impossible to obtain physically. Motivated by the desire to reduce the need to run repeated CFD simulations, we have in recent years demonstrated our successful use of CFD with z-transform, the discrete-time counterpart of the Laplace transform, to provide a simple, convenient, and robust method for simulating and analyzing the dynamics of mixing processes. Examples that we had illustrated included the open-loop response of a lumped-parameter mixing process (Yin and Yu, 2014 AIChE Annual Meeting Paper 354913), the open-loop response of a distributed-parameter mixing process (Yin, 2015 AIChE Annual Meeting Paper 404589), and the closed-loop response of an unstable mixing process (Yin, 2016 AIChE Annual Meeting Paper 449915).

While we may normally strive to design and implement mixing processes as ideal continuous stirred-tank reactors, the behavior of real mixing processes often does not conform to this paradigm. In order to run mixing processes more effectively and efficiently or to troubleshoot problematic ones, it is helpful to be able to characterize the internal mixing dynamic behavior of such nonideal processes in a quantitative and deterministic way. In our current study we extend our previously work by opening up the blackbox nonideal mixing process and demonstrate how we can analyze its internal mixing dynamics using transient CFD simulations and z-transform. The result allows us to identify and construct models to capture local phenomena such as compartmentalization and short-circuiting, and determine under which conditions the overall mixing process can be deconstructed into a network of smaller classical mixing and transport processes.

Reactive Mixing in a Stirred Tank Reactor

Mahsa Taghavi
Jafarsadegh Moghaddas

Sahand University of Technology
Tabriz, Iran

10/30/2017

1

Contents

- Introduction
- Experimental
- Results and discussion
- Conclusion

2

Introduction

Mixing: Reduction of inhomogeneity in order to achieve a desired process result

Mixing plays a key role in:

- Chemicals and Petrochemicals
- Biotechnology
- Polymer processing
- Food
- Pulp and Paper
- Drinking water and wastewater treatment

Stirred Tank: Physical movement of material between various parts of the tank using rotating impeller to give enhanced rates of mass transfer

3

Introduction

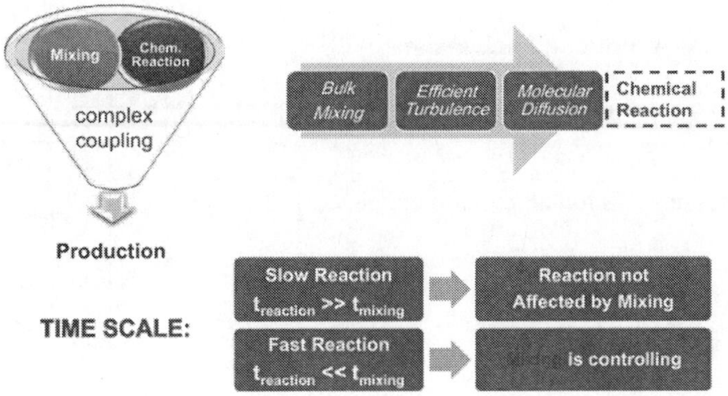

$$Da_M = \frac{t_M}{t_R} = \frac{Reaction\ rate}{Mixing\ rate}$$

4

Introduction

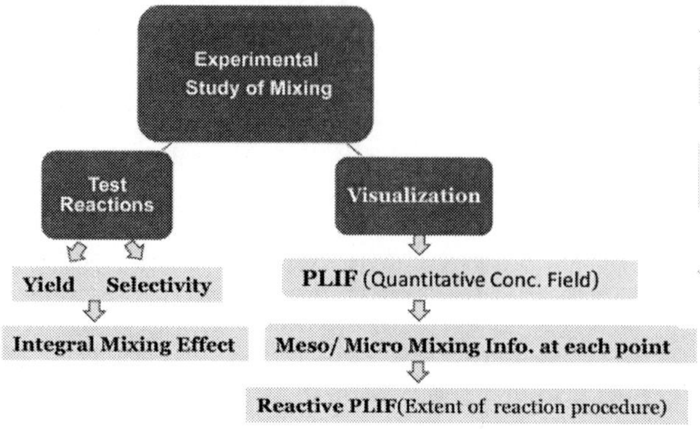

PLIF Technique Advantages:

Non-intrusive technology
Direct measurements
Use in reacting and non-reacting flows
Instantaneous visualization with high temporal and spatial resolution
Quantitative, precise and accurate measurements of concentration or temperature fields

5

Introduction

Principle of PLIF (Planar Laser Induced Fluorescence):

Excitation of molecules to higher levels of energy via laser absorption which causes light emission

The intensity of this fluorescence is a function of the concentration:

$I = \Phi . C$

I: intensity of light
Φ: constant
C: concentration

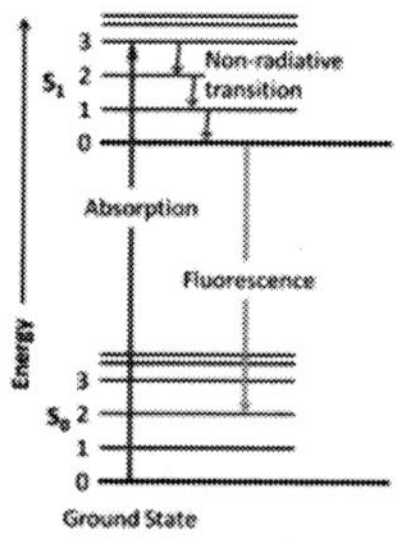

6

Introduction

Literature review on mixing syudy in stirred tanks using PLIF technique:

Researcher	Year	Reactive Mixing
Gaskey	1990	-
Andre	1992	-
Houcine	1996	-
Distelhoff	1997	-
Guillard	1999	-
Fall	2001	-
Hall	2004	-
Moghaddas	2004	-
Chung	2007	-
Yek	2009	-
Collignon	2010	-
Busciglioa	2012	-
Alvarez	2004	✓
Hu	2010,2012	✓

7

Experimental

Reactive PLIF Basic idea: Use a reaction to quench the fluorescence signal and quantitatively record mixing process

Oxidation Reaction: Fenton Reaction

Strong, Environmentally Friendly

$Fe^{2+} + H_2O_2 \rightarrow Fe^{3+} + HO^{\cdot} + HO^{-}$ | k= 10^2 L mol^{-1} s^{-1}

$HO^{\cdot} + RhB \rightarrow M$ | k= 10^7-10^8 L mol^{-1} s^{-1}

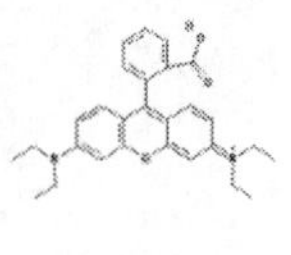

Rhodamine B

8

Experimental

- Laser: Pulsed Nd:YAG (532 nm)
- Sheet optics: Transform laser beam into a sheet
- Fluorescent dye: Rhd B
- Stirred tank reactor: Rushton turbine, D/T=1/3
- CCD Camera
- PC and image processing software: Post processing raw images for error correction and calibration

✓ Tank is filled with tracer and $(NH_4)_2Fe(SO_4)_2$

✓ H_2O_2 is injected to system with a digital pump

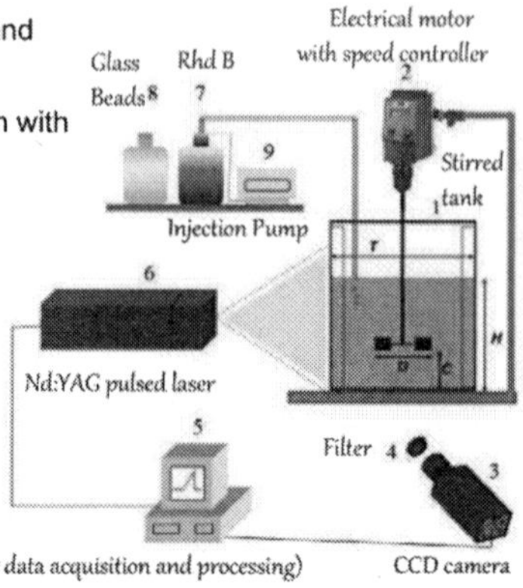

9

Results and discussion

Non-Reactive Mixing, N=100 rpm, C/T=1/3, Surface injection

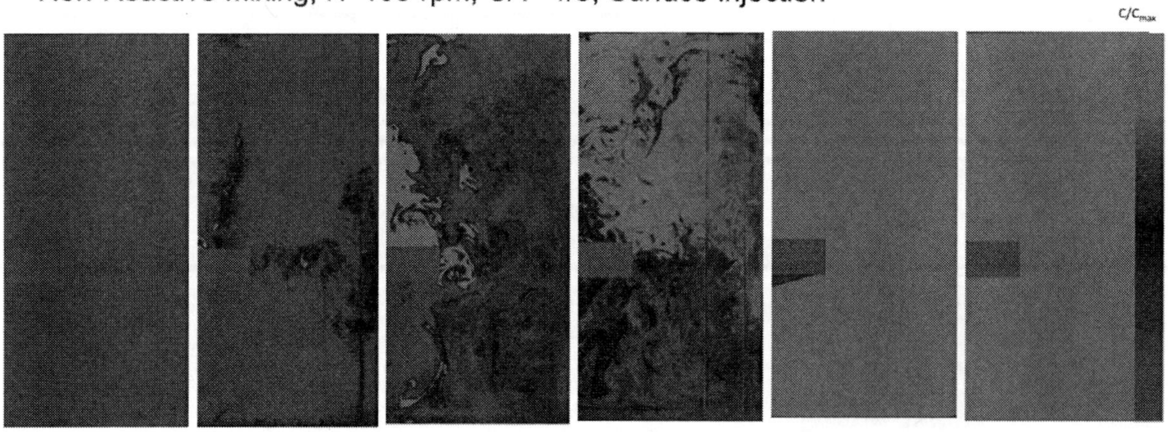

0 s 2 s 4 s 8 s 16 s 25.38 s

10

Results and discussion

Reactive Mixing, N=100 rpm, C/T=1/3, Surface injection

0 s   2 s   4 s   6 s   8 s   10.54 s

Results and discussion

Effect of impeller clearance on mixing and reaction times:
C/T=1/3, Surface injection

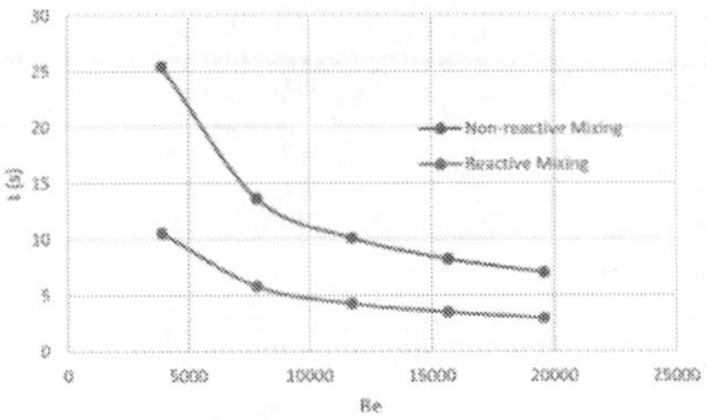

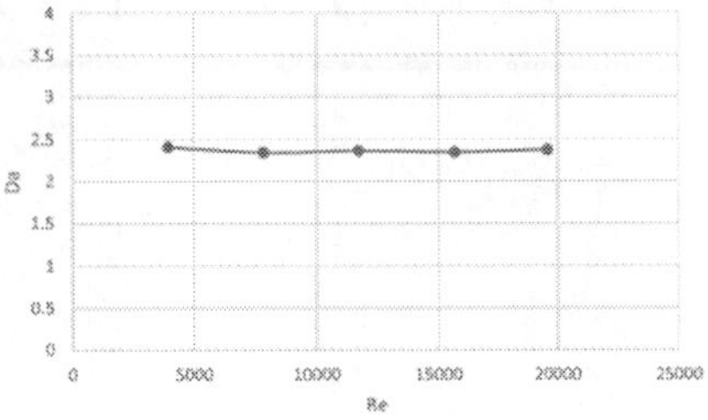

Results and discussion

Effect of impeller clearance on reaction time:

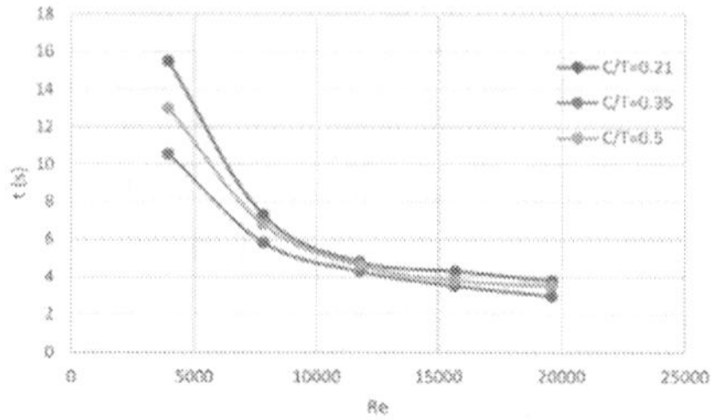

13

Results and discussion

Effect of injection location on reactant distribution in a reactive mixing:

Bottom injection | Impeller region injection | Surface injection

14

Conclusion

- Mixing time and reaction time in a Rushton turbine stirred reactor are successfully determined by PLIF technique.
- For Fenton oxidation reaction, reaction time is lower that mixing time under different mixing conditions and mixing controls the reaction rate.
- Damköhler number (Da) which is the ratio of reaction rate to mixing rate, is about 2.35 for this reaction under turbulent Re numbers.
- Optimizing the impeller clearance to C/T=1/3, leads to improving reactive mixing rate.
- Injecting the reactant in the impeller region leads to a better distribution of it throughout the reactor and homogeneity could be obtained in a shorter time.

15

Conclusion

- Under Da>1, improving mixing conditions would lead to increasing the reaction rate.
- Reactive PLIF technique shows details of concentration fluctuations in every point of the reactor, and so it is an effective technique in accurate recording of reactive flow behavior.

16

503670 Digital Alchemy for Assembly Engineeringc

Abstract:

Back in the day, Stuart Churchill was a pioneer in the use of computers for modeling chemical and thermal processes. Since that time, the role of computing in chemical engineering has become indispensable. In this talk, we present new computational methods we call "digital alchemy" that allow for predictive modeling of thermodynamic assembly, and permit the design of building blocks optimized for assembly into target structures with desired properties. In this approach, standard thermodynamic ensembles are generalized to include alchemical variables and their thermodynamic conjugates, allowing for rapid, real-time computational investigation of large, complex design spaces.

503671 Improved Algebraic, Numerical, and Graphical Representations in Fluid Mechanics

Abstract:

The concept of equivalents is applied to fully developed flows through straight channels, and two applications are introduced: the equivalents that produce equal rates of flow through channels of different cross-section, and the diminishment that occurs between the regimes of laminar and turbulent flow. The validity of the eddy diffusivity and the invalidity of the mixing-length as measures of the shear stress due to turbulence have perhaps never before been demonstrated so clearly. The possible choices of primary variables and the consequences are also examined. The objective of these explorations is an improved understanding of the basic fluid mechanics by teachers, students, and industrial practitioners.

503672 The Scaling of Turbulence Near the Wall and the Churchill Turbulent Flux Correlation: Insights with Lagrangian Simulations

Abstract:

The evaluation of mean velocity and mean temperature in wall-bounded turbulent flows has been based on the "classical" theory of turbulence and the law of the wall. The viscous wall parameters, namely the friction velocity, the friction temperature, and the kinematic viscosity of the fluid have been used to scale not only the mean values of velocity, but also other mean turbulent quantities close to the wall, including the Reynolds stresses and the turbulent velocity and temperature fluctuations. Stuart Churchill, as he questioned the validity of the Reynolds analogy between momentum and heat or mass transfer in his works [1], he realized that extending the scaling of momentum to the scaling of scalar quantities in turbulence was not a sound approach. He proposed a scaling of turbulence that depends on the turbulent momentum flux and the turbulent heat flux. In his AIChE Institute Lecture (in Miami, 1998), Churchill presented a summary of his work on this topic, and closed form solutions for the mean velocity and temperature [2]. The turbulent Prandtl number, the most significant parameter for the modeling of turbulent transport not only in theory but also in modern CFD commercial software packages, was also interpreted physically based on these ideas [3]. We have used direct numerical simulation results and results from Lagrangian simulations [4] for the transport of passive scalars in turbulent channel flows to explore Churchill's ideas on turbulent flow scaling and to develop an updated theory for the scaling of heat transport in turbulent flows [5-7]. His passion to remove empiricisms and approximations from the development of predictive correlations in turbulent transport has been an inspiration. In this talk, we will revisit the Churchill scaling in view of our simulation results, and will discuss the counterintuitive finding that wall turbulence can be used to separate rather than mix substances under special circumstances [8].

1. Churchill, S. W., 1996, "A Critique of Predictive and Correlative Models for Turbulent Flow and Convection," Ind. Eng. Chem. Res., 35, pp. 3122–3140.
2. Churchill, S.W., 2000, "Progress in the Thermal Sciences: AICHE Institute Lecture," AIChE Journal, 46(9), pp. 1704-1722.
3. Churchill, S. W., 2002, "A reinterpretation of the Turbulent Prandtl Number," Ind. Eng. Chem. Res., 41, pp. 6393–6401.
4. Papavassiliou, D.V, 2002, "Turbulent transport from continuous sources at the wall of a channel," Int. J. Heat Mass Transfer, 45(17), pp. 3571-3583.
5. Le, P.M., and Papavassiliou, D.V., 2008, "On the scaling of heat transfer using thermal flux gradients for fully developed turbulent channel and Couette flows," Int. Commun. Heat Mass Transfer, 35(4), pp. 404-412.
6. Srinivasan, C., and Papavassiliou, D. V., 2011, "Prediction of Turbulent Prandtl Number in Wall Flows With Lagrangian Simulations," Ind. Eng. Chem. Res., 50(15), pp. 8881–8891.
7. Srinivasan, C., and Papavassiliou, D.V., 2013, "Heat transfer scaling for wall bounded turbulent flows," Applied Mechanics Review 65(3), Art. 031002.

8. Nguyen, Q., Srinivasan, C., and Papavassiliou, D.V., 2015, "Flow induced separation in wall turbulence", Phys Rev E, 91, Art. 033019.

503674 Flow Boiling Using a Piranha Pin Fin Heat Sink

Abstract:

Flow Boiling using a Piranha Pin Fin Heat Sink

Corey Woodcock, Xiangfei Yu, Yoav Peles[1] and Joel Plawsky

Rensselaer Polytechnic Institute; [1]University of Central Florida

The *Piranha Pin Fin* (*PPF*) microdevice is a heat exchanger engineered to dissipate extreme thermal loads using either dielectric fluids such as HFE7000 or water. Fabrication of the device is CMOS compatible and the first generations of the device have been fashioned from silicon substrates. The PPF operates on the principle that heat transfer would be more efficient if heated fluids could be removed nearly as fast as they are created. This process would keep the driving force as high as possible. Thus the heat exchanger is built as two layers; a primary substrate that dissipates the bulk of the heat and shunts the heated fluids to a booster section that continues to heat the fluid and extends the total surface area of the device. Three generations of devices were fabricated. The first two generations were single-layer systems designed to help understand the operation of the primary layer and optimize the shape and location of the pin fins. The third generation was the first, two-layer system to be fabricated and tested.

Figure 1 Representative PPF array

Correlations for the heat transfer coefficient, h_{tp}, based on the Boiling Number and the Weber Number in the PPF design was developed and validated with Gen II. The discrepancy of heat transfer coefficient between predicted and measured values was within 20%. The ratio of two-phase to single-phase pressure drop was correlated with the Boiling Number (Bo) and the Jakob Number (Ja). The discrepancy of the two-phase multiplier between prediction and experimental measurements were within 12%.

The third generation *PPF* heat sinks were the first series to provide both a primary substrate populated with *PPFs*, as well as a booster substrate populated with solid micro pin fins. The substrates are micro-aligned and fusion-bonded to provide a full three dimensional (3D) architecture with bulk silicon properties. Experimental results from the Gen III *PPF* microdevices provide a proof of concept of the true *PPF* heat sink concept with fluid venting into multiple substrates. At low flowrates (G = 670 [kg/s/m^2]) the Gen III *PPF* heat sink has been shown to dissipate heat loads approaching 1.1 [kW/cm^2] at an exit quality of 95%, while at higher flowrates (G = 3,300 [kg/s/m^2]) the thermal limit of the heat sink remains unknown due to exceeding the available power supply at >1.4 [kW/cm^2].

503673 Transport Problems in the Spirit of Stuart Churchill for Teaching and Research at the University of Michigan

Abstract:

Stuart Churchillâ€™s solutions to important heat transport problems have greatly influence both pedagogy and research. The correlation of Churchill and Chu for the natural convection heat transfer to or from a vertical plate, over an astonishing range of Rayleigh numbers is based partly on a similarity solution of a type common in heat, mass, and momentum transfer for problems in chemical engineering. One such problem is asphaltene or wax transport and deposition onto the wall of an oil pipeline, which can be solved using similar methods, as will be discussed. In addition, such problems, and the wealth of their applications make for excellent motivations and examples in the teaching of graduate or undergraduate transport, as will be discussed in the talk.

503466 The Development of a Correlation for Solid-Liquid Cloud Height in Mechanical Agitated Vessels

Abstract:

Agitated vessels containing mixture of solids and liquids can be characterized by a minimum suspension speed and a vertical solids distribution. In many cases the cloud height which is the visual demarcation between slurry and clear interface is a useful parameter for characterizing vertical distribution. The emphasis in these studies has been on the effect of physical properties: concentration, density and viscosity and the effect of scale up. Geometry has been fixed throughout.

This correlation is of a dimensionless cloud height which is normalized using the tank diameter. This work presents additional data at high particle densities and weight fractions. Based on this work it was found that the dimensionless cloud height as a function of solids weight fraction has two regimes. As the weight fraction is increased at a constant stirrer speed the cloud height initially shows a slight increase reaching a local maximum and then exhibits a drop in height. Based on this observation the cloud height correlation has been modified to predict this behavior.

The variables with the largest effect on cloud height are the equipment variables of impeller speed and diameter. The physical properties with the largest effect on cloud height are density and particle diameter.

493633 Impeller Power Draw during Turbulent Operation in Liquid-Solid Suspensions

Abstract:

Impeller power draw in turbulent liquid-only operation has been studied extensively, while in multiphase operation, power draw during gas dispersion is relatively well understood. Far less work has been performed regarding impeller power draw when agitating liquid-solid suspensions with settling solids. Additionally, most work in liquid-solid systems has been limited in scope, focusing on a particular impeller type with a narrow range of particle characteristics and loadings.

The most common approach to predicting impeller power draw in liquid-solid systems is modification of the liquid-only power number to account for the effect of solids on mixture density. Typical of this approach is the viewpoint of Nagata (1975) that power draw can be estimated using the average density in the vicinity of the impeller, although this density is difficult to determine. Due to this complication, the average suspension density is often used in practice, but Barresi and Baldi (2000) reported that use of the average suspension density generally over estimates impeller power draw. Chudacek (1982) postulated that use of the suspension density approach requires proper determination of the suspended solids mass and the volume through which they're distributed. Thus, unsuspended solids should not contribute to the solids mass and also reduce the suspension volume. Similarly, clear liquid above the suspended solids should not be included in the suspension volume. Rather than using slurry density, Micheletti et al. (2003) recommended using the liquid-only power number and liquid density with a correction factor based on the difference between the average slurry and liquid densities.

In an effort to develop a better understanding of impeller power draw during turbulent operation in liquid-solid systems, an experimental study was performed using six impeller types: a narrow-blade hydrofoil, a wide-blade hydrofoil, a pitched-bladed turbine, a straight-blade turbine, a disc turbine with straight blades, and a disc turbine with semicircular blades. To reduce solid distribution effects, data was taken only when solids were distributed throughout the liquid phase. A variety of solids were studied across a wide range of loadings, yielding significant density variations. Data was also taken with neutrally buoyant particles whose addition did not change suspension density.

This study indicates that the six impellers exhibit different behaviors with respect to the effect of suspension density on turbulent impeller power draw. For some impellers, the addition of solids leads to power draw increases that are less than the increases in suspension density. For other impellers, the increases in power draw are similar to the increases in suspension density. And in one case, the increases in power draw are larger than the increases in suspension density. The correlating approach of Micheletti et al. (2013) is found to be promising for all impellers, but again, each impeller behaves differently.

References

Barresi, A. A., and G. Baldi, "Power Consumption in Slurry Systems", *Proceedings of the Tenth European Conference on Mixing*, pages 133-140 (2000).

Chudacek, M. W., "Formation of Unsuspended Solids Profile in a Slurry Mixing Vessel", *Proceedings of the Fourth European Conference on Mixing*, pages 275-287 (1982).

Micheletti, M., L. Nikiforaki, K. C. Lee, and M. Yianneskis, "Particle Concentration and Mixing Characteristics of Moderate-to-Dense Solid-Liquid Suspensions", *Industrial and Engineering Chemistry Research*, Volume 42, pages 6236-6249 (2003).

Nagata, S., *Mixing: Principles and Practice*, Kodansha Ltd, Tokyo / John Wiley and Sons, New York (1975).

495420 Novel Experimental Method for the Determination of the Minimum Agitation Speed for Solids Suspension in Flat-Bottomed Stirred Tank Reactors

Abstract:

Knowledge of the minimum agitation speed, N_{js}, required to suspend finely divided solids in vessels stirred by an impeller is a critical parameter to properly operate industrial tanks in a large number of industrial operations. The most common experimental approach to measure N_{js} is that of Zwietering's (*Chem. Eng. Sci.*, 1958, **8,** 244-253), consisting of visually inspecting the tank bottom and visually determining the impeller agitation speed at which the solids are observed to rest on the tank bottom for no more than 1-2 seconds before being swept away. This method is quite reliable, but a method not relying on the operator would clearly be preferred. Therefore, the main objective of this work was to develop a simple, repeatable, and observer-independent method to determine N_{js} for a variety of agitation systems and agitation conditions in tanks with a flat-bottom, the most common tank shape for which N_{js} has been obtained in previous studies. Flat bottom reactors are also among the most widely used reactors for research and development studies in the pharmaceutical and biopharmaceutical industry.

In this study, N_{js} was experimentally determined by using a newly developed method in which the area covered by the unsuspended solids still at the bottom of the tank was measured at increasing agitation speeds, starting at low speeds (below N_{js}). To do so, images of the tank bottom were captured in .jpg format by a digital camera. Each image was processed with the appropriate software (Image J) to quantitatively determine the area still covered by solids at that speed. Increasing the agitation speed increased the amount of solids being suspended, resulting in a decrease in the area covered by solids at the bottom of the tank. Plots of the area covered by the solids vs. the corresponding agitation speed resulted in a linear function, which when extrapolated to A®0 yielded the expected value of N_{js} (named $N_{js\text{-}A}$). The values so obtained for $N_{js\text{-}A}$ were then compared to the N_{js} value determined visually ($N_{js\text{-}vis}$).

This approach was tested for a number of mixing systems with different impeller types (disk turbines, flat-blade turbine, pitched-baled turbines, hydrofoil impeller), impeller off-bottom clearances, impeller sizes, and tanks sizes. Preliminary tests were conducted using the same method but in hemispherical tanks, where the solids at the bottom of the tank a formed an approximate circular pattern. As compared to hemispherical bottomed tank, the solid particles in flat-bottomed tanks were dispersed all over the bottom of the tank. Therefore, the sum of all the individual areas covered by the solid particles deposited at the bottom was quantitatively determined. The results for $N_{js\text{-}A}$ obtained with the new method were compared with those obtained using the traditional Zwietering's approach. In general, excellent agreement was found between $N_{js\text{-}A}$ and $N_{js\text{-}vis}$. It can be concluded that this newly developed method constitutes a novel, reliable, and, especially, *observer-independent* method to experimentally determine N_{js}.

It is expected that this approach could be of particular relevance in a variety of mixing applications, including the suspension of micro-carrier particles, such as glass beads, on to which animal cells can become attached and used to produce various bio-products, and in the suspension of beads for production of proteins.

499591 Simulating Solid Suspension in Stirred Vessels with a Fully Coupled CFD-DEM Algorithm

Abstract:

Solid – Liquid mixing in high viscosity liquid and/or high solids concentration is a challenging problem due to the solid-solid and solid-liquid phase interactions. Just suspended speed (N_{js}), suspension height (H_s) and the power consumption can vary not only with the material properties but also with the design of tank and impellers.

At the AIChE annual meeting 2016 we presented a new method for simulating solid suspension based on a fully coupled CFD-DEM algorithm [1]. Validation was done against experimental data for a broad variety for impeller speed.

In this contribution we are evaluating the effect of material properties and different bottom clearance on the fraction of suspended particles and the occurrence of dead zones below the impeller. Additionally we are comparing the achieved results with results based on an Eulerian Multiphase – Granular Flow (EMP-GF) approach with respect to accuracy and runtime. The differences and limitations of both approaches are discusses.

[1] Baran, O. et al.: A Fully Coupled CFD-DEM Algorithm for S-L Mixing in Stirred Vessels, AIChE Annual Meeting 2016, San Francisco, Ca

[2] Bertrand, F. et al.: Experimental investigation of the mixing of viscous liquids and non-dilute concentrations of particles in a stirred tank, Chem. Eng. Res. Des. (108), 2016, 55–68

499786 Understanding Particle Attrition in Agitated 3-Phase Aerated Slurry Reactors

Abstract:

In 3-phase slurry reactors, a solid phase catalyst is used to promote a reactor between a liquid phase component and a gas phase component. Since the reaction requires intimate contact between all 3 phases, the system must be vigorously agitated in order to adequately suspend the solid particles and produce small gas bubble to enhance mass transfer. However, these high energy dissipation rates are detrimental to the catalyst particles.

To understand the impact of process variables on catalyst attrition, a study was performed where a series of experiments, aided by computational fluid dynamics simulations, were undertaken. This study helped to elucidate the type of attrition and allowed a first-order estimate of catalyst lifetime.

495479 A Unifying Framework for Mass Transfer Dynamics in Gas â€“ Liquid Segmented Flow in a Circular Tube

Abstract:

Knowledge of the parameters that characterize mass transfer and reaction dynamics in gas-liquid systems is critical for the design of macroscale units for separation and reaction processes. To unearth these parameters, researchers have been replicating these processes on a microscale. The advantages of this approach are the lack of turbulence, availability of high interfacial areas and well-characterized flow profiles. Our work focuses on one such category of the flow profiles - the Taylor flow. Taylor flow finds application in design of monolithic catalytic reactors and foam flow through porous media in enhanced oil recovery processes. Recently, it has been used on microfluidic platforms to determine thermodynamic and kinetic parameters of special gas-liquid systems such as CO_2 and switchable solvents. Generation of alternating segments of gas and liquid, and subsequent observation of the gas bubble lengths upon dissolution under different conditions has the capability of supplying these system parameters by using appropriate theoretical models. Literature presents multiple models for data deconvolution, but they pose three major limitations. First, mass transfer between the bulk liquid segment and the liquid film above it has been incorrectly estimated. Second, an average mass transfer coefficient is assumed to be valid throughout the dissolution process. Third, the liquid segment is assumed to be well mixed, even though there is clear evidence to the contrary in the literature.

In this work, we have rectified all of the above limitations to arrive at a unifying framework. A scaling analysis was conducted to delineate the separate contributions of the liquid films above the gas bubble and the liquid segment to mass transfer. We identified four operational regimes governed by the dimensionless bubble length, dimensionless liquid segment length, the Peclet number and the capillary number. We observed regime cross-overs, in addition to a decrease in the mass transfer coefficient by nearly a factor of three upon bubble dissolution. We also included the effect of finite mass transfer time across the liquid segments owing to its closed streamline nature. We have delineated the parameter regimes where past models are special cases of our current model, thus identifying where these models are likely to fail. To validate our analysis approach, experiments were carried out in circular, silica capillaries of different radii by generating segmented flow of CO_2 in physical solvents such as ethanol, acetonitrile and propylene carbonate. The results presented in this work will provide a framework to guide the design of experiments that will accurately determine thermodynamic and kinetic parameters relevant to gas-liquid systems. Moreover, the model can predict mass transfer dynamics even in non-dissolving systems, where Taylor flow is used solely for a controlled residence time distribution. Finally, it finds application in the design of catalyst-coated microchannel contactors, which is a step towards designing gas-liquid-solid reactions in such setups.

497571 Phase Inversion of a Solid-Stabilized Emulsion: Effect of Particle Concentration

Abstract:

Phase inversion of solid stabilized emulsion could be a potential workaround for the commonly encountered problems in the desalting of crude oil by making water the continuous phase during the processing. In the present work, the effect of particle concentration on the catastrophic phase inversion point was investigated. The phase inversion was systematically triggered by the continuous addition of dispersed phase. During phase inversion, we focused on tracking the droplet morphology evolution by a Particle Vision Microscope (Mettler Teledo), which proved to be a powerful tool to monitor the instable process. The phase inversion was found to occur in a regime where the particles are insufficient to fully cover the oil-water interface. A linear relationship between the particle concentration and the phase inversion point was obtained. Our findings highlighted the fact that the phase inversion always takes place above a critical value expressed as the number of particle per surface area. This always occurred, regardless of the particle concentration. The systematic observation has been explained by relating the particle coverage to the coalescence rate of the system. It has been compared to pure transient coalescence experiments using sudden reduction of impeller speed, without particles.

JOHANNES KEPLER UNIVERSITY LINZ

A Eulerian-Eulerian-Lagrangian Hybrid Model for the Simulation of the Droplet Size Distribution of Liquid-Liquid Emulsions in Stirred Tank Reactors

Reza Farzad, Stefan Pirker and Simon Schneiderbauer

AIChE Annual Meeting 2017
Minneapolis, October 30th, 2017

CD Laboratory for Multi-Scale Modelling of Multiphase Processes
Department of Particulate Flow Modelling
Johannes Kepler University | Linz | Austria

Stirred Tank and mixing operation

Stirred tank reactors are considerably important in process industries such as pharmaceutical, food, polymer, oil and gas. It can handle different mixing operations:

Suspension
(solid-liquid)

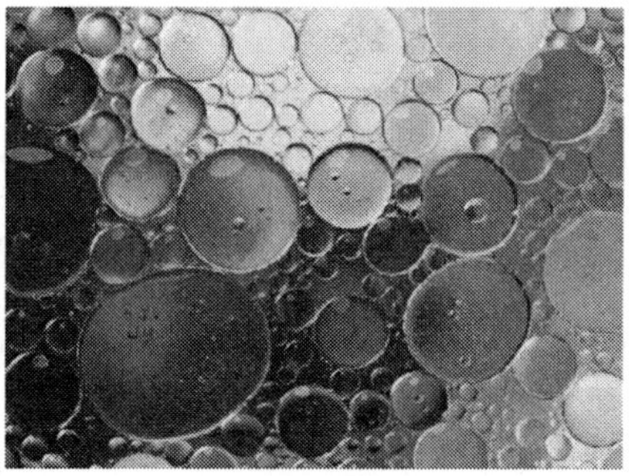

Emulsion
(liquid-liquid)

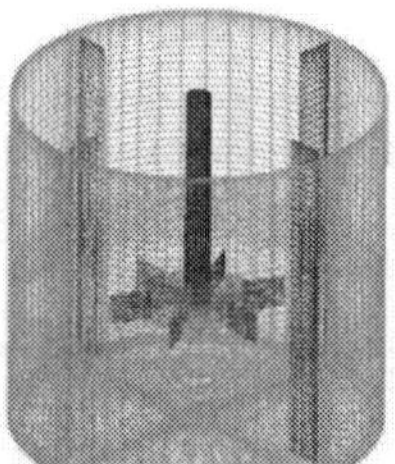

Sparging
(gas-liquid)

Drop size distribution (DSD)

Drop size distribution (DSD) plays a key role as it strongly affects the overall mass and heat transfer in the liquid-liquid systems.

w/o

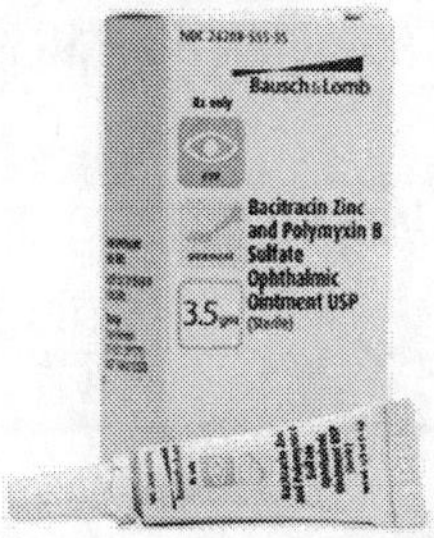

o/w (w/o)

o/w

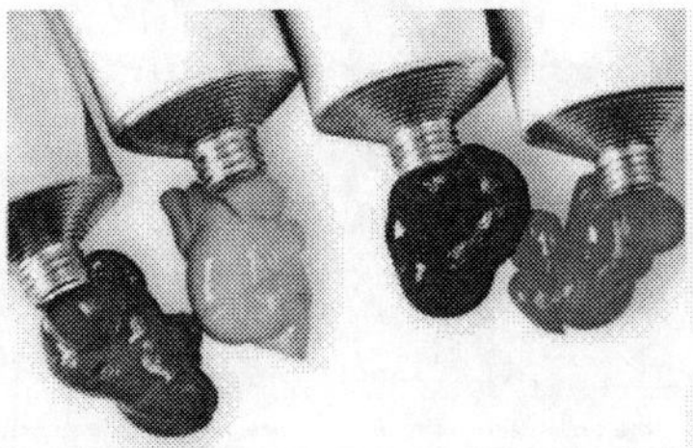

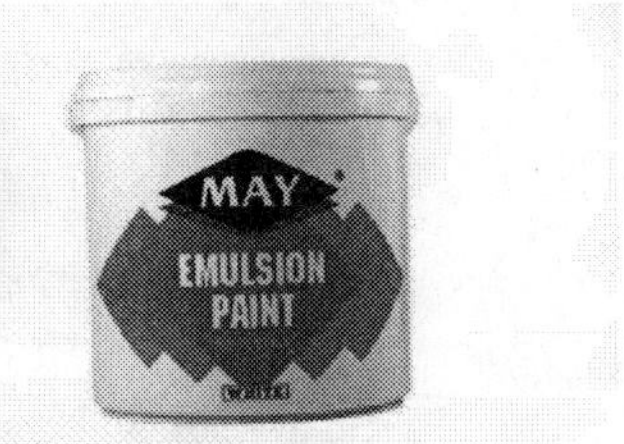

CFD simulation- Emulsions in STR Fundaments

Eulerian-Eulerian/Two Fluid model (TFM)

- **Resolving all droplets is not feasible computationally (Lagrangian approach)**
- **TFM treats liquid-liquid emulsion as interpenetrating continua (Eulerian-Eulerian)**

Continuous phase continuity: $\frac{\partial}{\partial t}(\alpha_d \rho_d) + \nabla.(\alpha_d \rho_d \boldsymbol{u}_d) = 0$

Continuous phase momentum:

$$\frac{\partial}{\partial t}(\alpha_d \rho_d \boldsymbol{u}_d) + \nabla.(\alpha_d \rho_d \boldsymbol{u}_d \boldsymbol{u}_d) = -\alpha_d \nabla p + \nabla.(\alpha_d \tau_d) + \boldsymbol{F}_d + \boldsymbol{F}_{t.d,c} + \alpha_d \rho_d \boldsymbol{g}$$

$\boldsymbol{F}_{t.d,c}$ **Turbulent dispersion force (Simonin and Viollet 1990)**

$\boldsymbol{F}_d = K_{cd}(\boldsymbol{u}_c - \boldsymbol{u}_d)$ **Drag force is considerably affected by Droplet Size Distribution**

How to obtain droplet size distribution?

CFD simulation- Emulsions in STR Fundaments

Hybrid TFM-DPM

- **Additionally track statistically representative Lagrangian tracer trajectories**
- **These trajectory "follow" the the dispersed phase by experiencing a "swarm" force**

Momentum exchange for parcel trajectory:

$$m_{d,k}\frac{\partial}{\partial t}(\boldsymbol{u_{d,k}}) = \frac{3}{4}\frac{\mu_c C_D Re}{\rho_c d_{d,k}^2} m_{d,k}(\boldsymbol{u_c} - \boldsymbol{u_{d,k}}) + \frac{C_D Re}{\tau_{dr.sw.k}} m_{d,k}(u_{d,k} - u_d) + \frac{\pi}{6} d_{d,k}^3(\rho_d - \rho_c)\boldsymbol{g}$$

Drag force on parcel — Droplets swarm force — Buoyancy force on parcel

Droplets swarm time scale: $\frac{1}{\tau_{dr.sw.k}} = \frac{3}{4}\frac{\mu_d \alpha_d}{\rho_d d_{d,k}^2}(1 - \alpha_d^{\frac{1}{3}})^{-1}$

Trajectory of a single Lagrangian tracer droplet

the continuous droplets move with locally averaged velocity

the impact of the collisions of the Lagrangian trajectory with the secondary phase

*Schneiderbauer, S. et al., 2017. A Lagrangian-Eulerian Hybrid Model for the Simulation of Poly-disperse Fluidized Beds: Application to Industrial-scale Olefin Polymerization. Powder Technology, 316, pp.697–710..

CFD simulation- Emulsions in STR Fundaments

- **The TFM is influenced by the Lagrangian diameter field:**

$$\langle d_{32}(\vec{x})\rangle = \frac{\sum_{p\in D} d_{d,p}^3 \delta(\vec{x} - \vec{x}_{d,p})}{\sum_{p\in D} d_{d,p}^2\, \delta(\vec{x} - \vec{x}_{d,p})} \qquad \delta(\vec{x} - \vec{x}_{d,p}) = \begin{cases} 0, & |\vec{x} - \vec{x}_{d,p}| > \frac{\Delta}{2} \\ 1, & |\vec{x} - \vec{x}_{d,p}| < \frac{\Delta}{2} \end{cases}$$

- **Liquid-liquid drag force:**

$$\boldsymbol{F}_d = K_{cd}(\boldsymbol{u}_c - \boldsymbol{u}_d) \qquad K_{cd} = \frac{\rho_d \alpha_d \alpha_c f}{\tau_{rel.}} \qquad f = \frac{C_D Re}{24}$$

$$\tau_{rel.} = \frac{\rho_d \langle d_{32}\rangle^2}{18\mu_d} \qquad Re = \frac{\rho_c |\boldsymbol{u}_c - \boldsymbol{u}_d| \langle d_{32}\rangle}{\mu_c}$$

Drop Size Distribution (DSD) in Taylor-Couette flow Theory

How to model breakup?

First we studied DSD in Taylor Couette Flows*

- **well-defined flow pattern**
- **Less complex flow vs. STR**
- **Most of the correlations provides global Sauter mean diameter as they are developed based on the global parameters.**

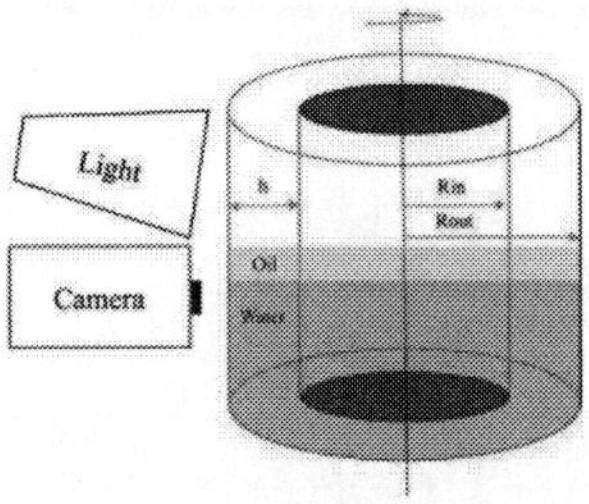

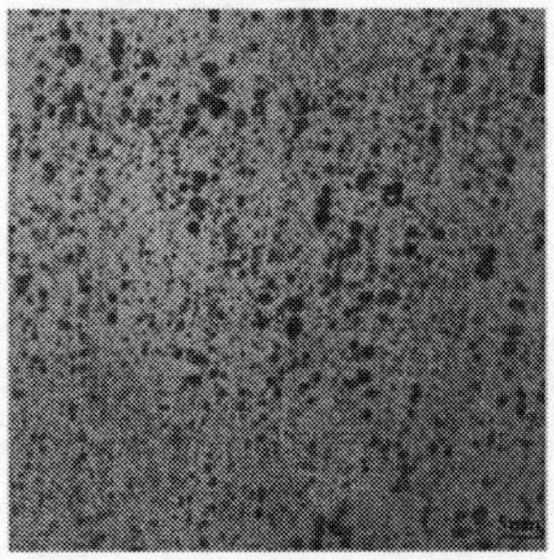

*Farzad, R., Puttinger, S., Pirker, S., & Schneiderbauer, S. (2017). Investigation of Droplet Size Distribution for Liquid-Liquid Emulsions in Taylor-Couette Flows. *Journal of Dispersion Science and Technology*

Drop Size Distribution (DSD) in Taylor-Couette flow Droplet breakup

$$Surface\ energy: 4\pi D^2\sigma \quad \textbf{Vs.} \quad Turbulent\ energy: \frac{4}{3}\pi D^3\rho\varepsilon$$

The probability of existence of a droplet with diameter of D is: $\epsilon = \int_{1/D}^{\infty} E(k)dk$

$$P(D) = \phi\left(\frac{\frac{4}{3}\pi D^3\rho \int_{1/D}^{\infty} E(k)dk}{4\pi D^2\sigma}\right)$$

$$E(k) = \alpha\varepsilon^{2/3}k^{-5/3} \qquad \int_{1/D}^{\infty} E(k)dk = \frac{3}{2}\alpha\varepsilon^{2/3}k^{2/3}$$

$$P(D) = \left(A\left(\frac{\rho_c}{\gamma}\right)^{\frac{3}{5}}\bar{\varepsilon}^{\frac{2}{5}}D\right)$$

Based on the Prandtl mixing length:

$$\bar{\varepsilon} = cv_t\dot{\gamma}^2 = cl_m{}^2\dot{\gamma}^3 \qquad We = \frac{\rho_c\dot{\gamma}^2 l_m{}^3}{\gamma}$$

$$D_{32} = \frac{\int_0^{\infty} D^3P(D)\,dD}{\int_0^{\infty} D^2P(D)\,dD}$$

Chen, Hsiao Tsung, and Stanley Middleman. "Drop size distribution in agitated liquid-liquid systems." *AIChE Journal* 13.5 (1967): 989-995.

Drop Size Distribution (DSD) in Taylor-Couette flow Theory

Inviscid fluid:

$$\frac{D_{32}}{l_m} = A\, We^{-\frac{3}{5}}$$

Viscous fluid: $\frac{D_{32}}{D_0} = (1 + cVi)^{\frac{3}{5}}$ *

$$Vi = \left(\frac{\rho_c}{\rho_d}\right)^{\frac{1}{2}} \mu_d (l_m{}^2 \dot{\gamma}^3)^{\frac{1}{3}} \frac{D_{32}^{\frac{1}{3}}}{\gamma}$$

$$\frac{D_{32}}{l_m} = A\, We^{-\frac{3}{5}} \left(1 + B \left(\frac{\rho_c}{\rho_d}\right)^{\frac{1}{2}} \frac{\mu_d}{\gamma} \left(D_{32} l_m{}^2 \dot{\gamma}^3\right)^{\frac{1}{3}}\right)^{\frac{3}{5}}$$

From the experimental data:**

$$\frac{D_{32}}{l_m} = 0.1\, We^{-\frac{3}{5}}$$

$$\frac{D_{32}^{(v)}}{l_m} = 0.08\, We^{-\frac{3}{5}} \left(1 + 9.9 \left(\frac{\rho_c}{\rho_d}\right)^{\frac{1}{2}} \frac{\mu_d}{\gamma} \left(D_{32} l_m{}^2 \dot{\gamma}^3\right)^{\frac{1}{3}}\right)^{\frac{3}{5}}$$

*Calabrese R V., Chang T P K and Dang P T 1986 Drop breakup in turbulent stirred tank contactors. Part I: Effect of dispersed phase viscosity. *AIChE J.* 32 657–66

**Farzad, R., Puttinger, S., Pirker, S., & Schneiderbauer, S. (2017). Investigation of Droplet Size Distribution for Liquid-Liquid Emulsions in Taylor-Couette Flows. *Journal of Dispersion Science and Technology*

Drop Size Distribution (DSD) in Taylor-Couette flow

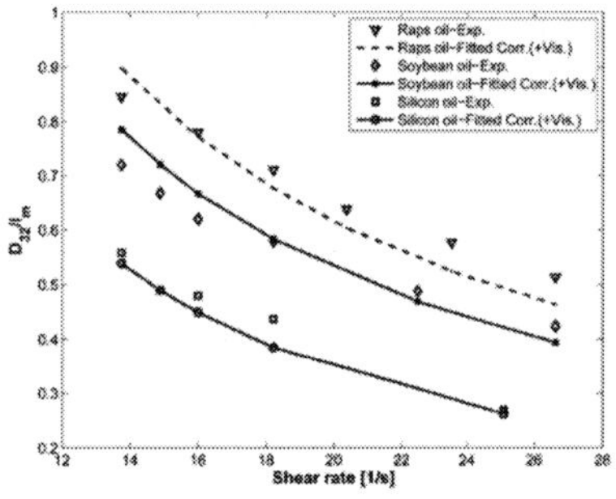

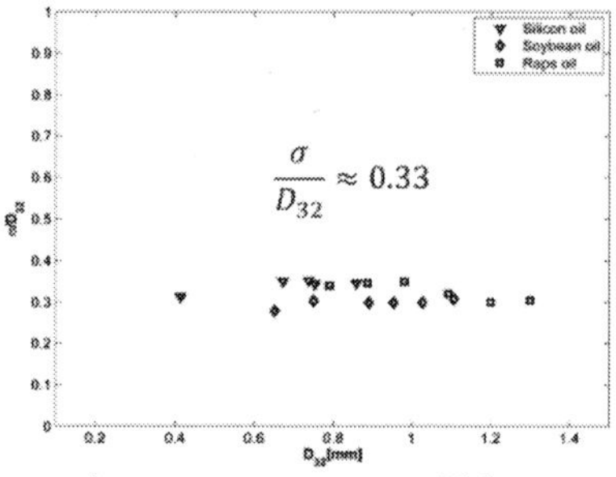

$$\frac{D_{32}}{l_m} = 0.08\, We^{-\frac{3}{5}} \left(1 + 9.9 \left(\frac{\rho_c}{\rho_d}\right)^{\frac{1}{2}} \frac{\mu_d}{\gamma} \left(D_{32} l_m{}^2 \dot{\gamma}^3\right)^{\frac{1}{3}}\right)^{\frac{3}{5}}$$

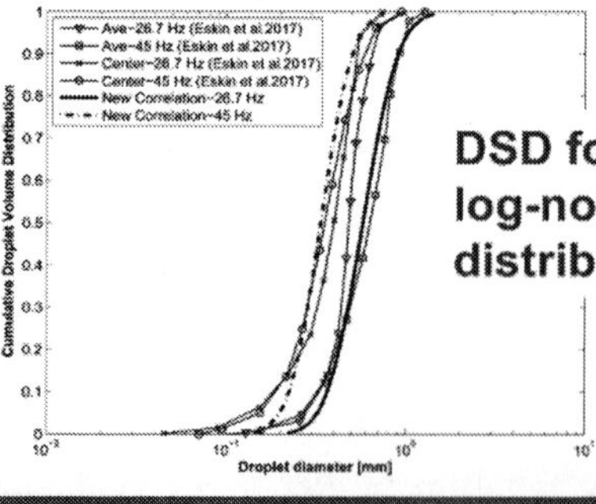

DSD follows log-normal distribution

*Farzad, R., Puttinger, S., Pirker, S., & Schneiderbauer, S. (2017). Investigation of Droplet Size Distribution for Liquid-Liquid Emulsions in Taylor-Couette Flows. *Journal of Dispersion Science and Technology*

Eskin, D., Taylor, S. D., & Yang, D. (2017). Modeling of droplet dispersion in a turbulent Taylor–Couette flow. *Chemical Engineering Science, 161*, 36-47.

CFD simulation- Emulsions in STR Fundaments

Novel Breakup "sub-grid" model

- Define an upper limit like D99 for each cell based on the correlation (log-normal)
- Droplet diameter > D99 ➡ it Breaks randomly into 2 smaller droplets

$$D_p = \exp(m + Z\,s)$$

$$m = \log\left(\frac{\mu^2}{\sqrt{\mu^2 + \sigma^2}}\right) \qquad s = \sqrt{\log\left(1 + \frac{\sigma^2}{\mu^2}\right)}$$

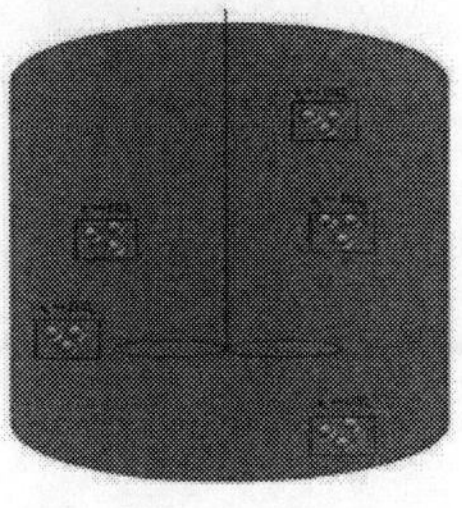

- We assumed: $\mu = D_{32}$ $\qquad \sigma = \frac{1}{3} D_{32}$

- The size of the daughter droplets can be calculated by selecting random Z by Box-Muller transform.

CFD simulation- Emulsions in STR Validation Cases

Boxall et al.

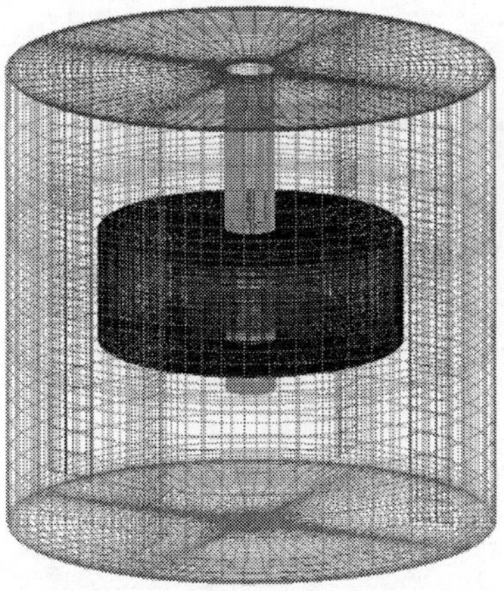

Case	Density (Kg/m^3)	Viscosity (kg/m.s)	Interfacial tension (mN/m)	Volume fraction
Water-in- Conroe oil (Boxall et al., 2010)	842	0.0031	20	15%
Water-in-Troika oil (Boxall et al., 2010)	869	0.02	11	15%

Boxall, J.A., Koh, C.A., Sloan, E.D., Sum, A.K., Wu, D.T., 2010. Measurement and Calibration of Droplet Size Distributions in Water-in-Oil Emulsions by Particle Video Microscope and a Focused Beam Reflectance Method. Ind. Eng. Chem. Res. 49, 1412–1418. doi:10.1021/ie901228e

CFD simulation- Emulsions in STR Results

Results of Breakup (coalescence is neglected) : Water-in-Conroe Oil

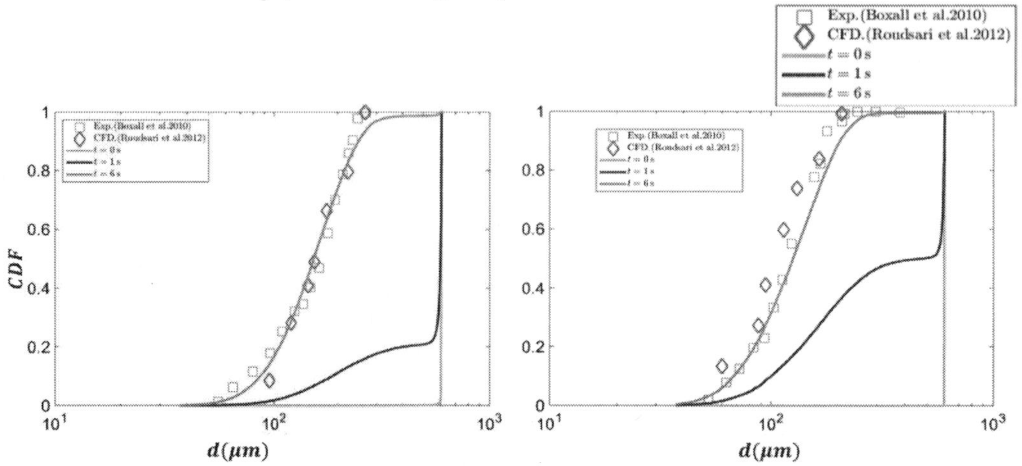

300 RPM

400 RPM

Roudsari, S. F., Turcotte, G., Dhib, R., & Ein-Mozaffari, F. (2012). CFD modeling of the mixing of water in oil emulsions. *Computers & Chemical Engineering, 45*, 124-136

CFD simulation- Emulsions in STR Results

Results of Breakup (coalescence is neglected) : Water-in-Conroe Oil

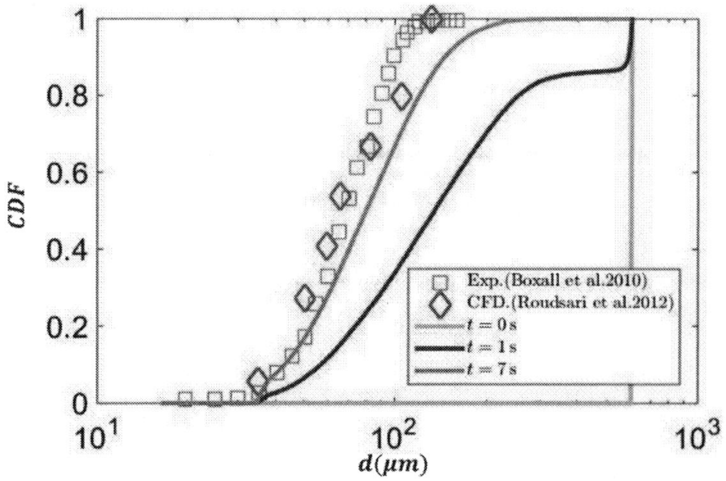

600 RPM

Roudsari, S. F., Turcotte, G., Dhib, R., & Ein-Mozaffari, F. (2012). CFD modeling of the mixing of water in oil emulsions. *Computers & Chemical Engineering, 45*, 124-136

CFD simulation- Emulsions in STR
Validation Cases

Case	Density (Kg/m³)	Viscosity (kg/m.s)	Interfacial tension (mN/m)	Volume fraction
Kerosene-in Water (Lovick et al., 2005)	828	0.0055	44.37	10%
Kerosene-in Water (Lovick et al., 2005)	828	0.0055	44.37	20%

- **Equilibrium D32(Vol%)= F(Vol%,D32(1%)) in order to neglect a term for coalescence**

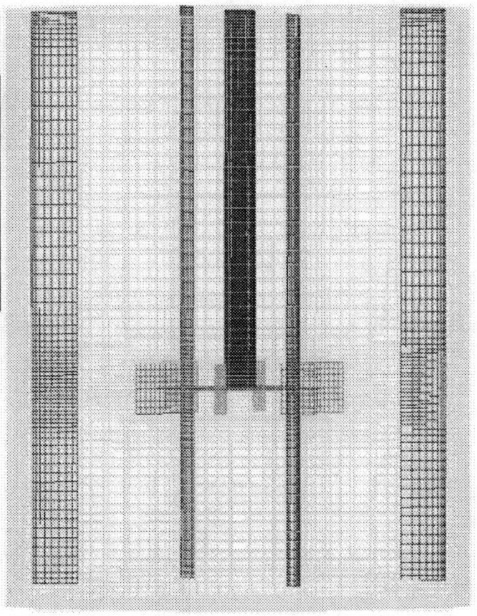

Lovick, J., Mouza, A. A., Paras, S. V., Lye, G. J., & Angeli, P. (2005). Drop size distribution in highly concentrated liquid–liquid dispersions using a light back scattering method. *Journal of Chemical Technology and Biotechnology, 80*(5), 545-552.

CFD simulation- Emulsions in STR
Results

Results of Breakup (coalescence is neglected) : Water-in-Troika Oil

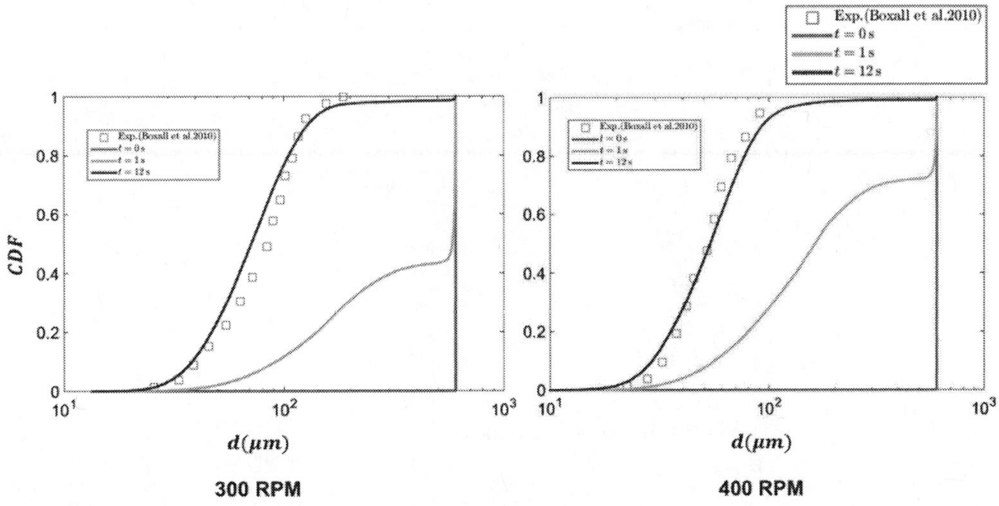

CFD simulation- Emulsions in STR Results

Results of Breakup (coalescence is neglected) : Kerosene-in- Water

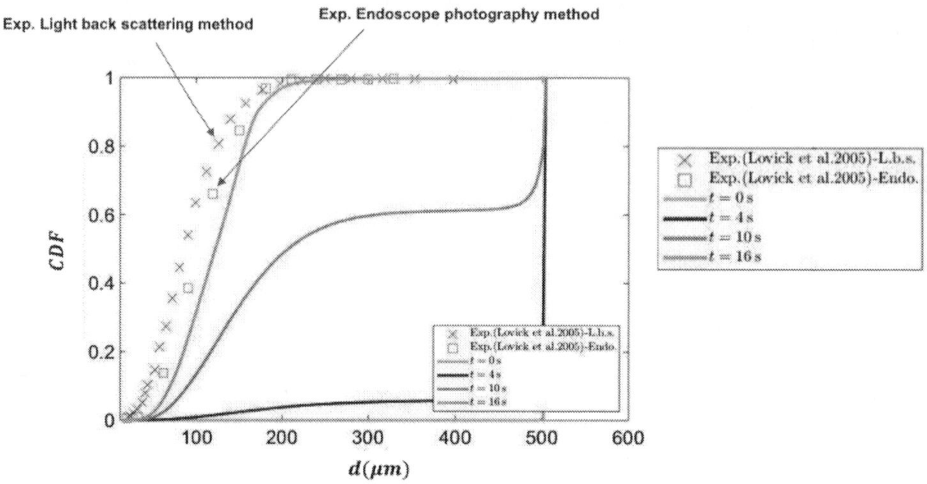

10% Kerosene-in Water at 450 RPM

Lovick, J., Mouza, A. A., Paras, S. V., Lye, G. J., & Angeli, P. (2005). Drop size distribution in highly concentrated liquid–liquid dispersions using a light back scattering method. *Journal of Chemical Technology and Biotechnology, 80*(5), 545-552.

CFD simulation- Emulsions in STR Results

Results of Breakup (coalescence is neglected) : Kerosene-in- Water

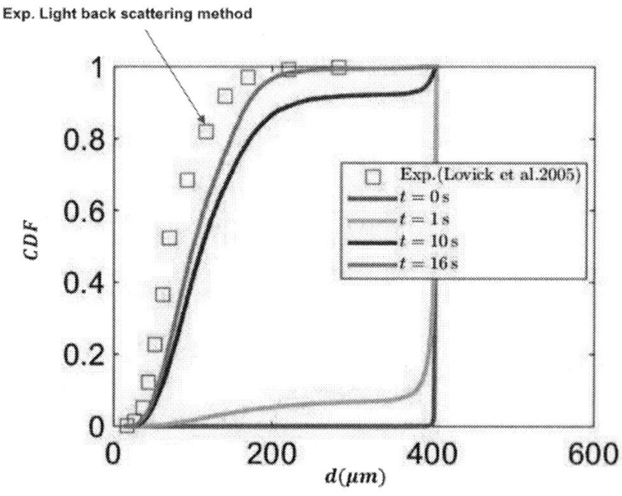

20% Kerosene-in Water at 550 RPM

Conclusion

Conclusion

- Breakup model works fairly good for liquid-liquid cases and the results are in a good agreement with the experimental data form the literatures.
- The Lagrangian information such as droplet resident time can be obtained directly.
- The number of droplets in a real process can be represented.

OUTLOOK

- Account for droplet coalescence directly

JOHANNES KEPLER UNIVERSITY LINZ

Thank you for your attention!

CD Laboratory for Multi-Scale Modelling of Multiphase Processes
Department of Particulate Flow Modelling
Johannes Kepler University | Linz | Austria

CFD simulation- Emulsions in STR Fundaments

Eulerian-Eulerian (TFM)+Population balance equation (PBE)

$$\frac{\partial}{\partial t}(\rho_2 n_{di}) + \nabla.(\rho_2 n_{di}\widetilde{u_2}) = \rho_2[B_B - D_B + B_C - D_C]$$

- **PBE commonly is used to simulate the DSD in the particulate flows.**
- **The poly-dispersed particle/droplet drag issue: Smaller particles travel faster than the larger ones .**
- **Considering a separate momentum equation for each size class increases the computational cost.**

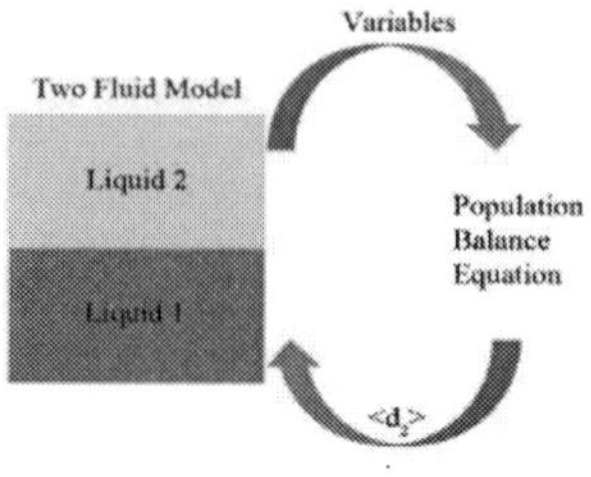

CFD simulation- Emulsions in STR Fundaments

Breakup: Numerical treatment

$$D_p < d_{d,k}$$

A) 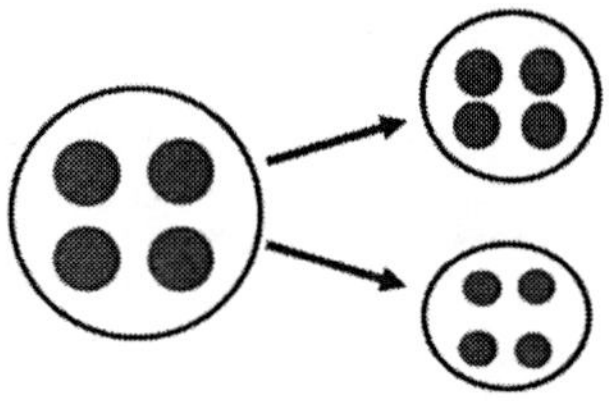

B) 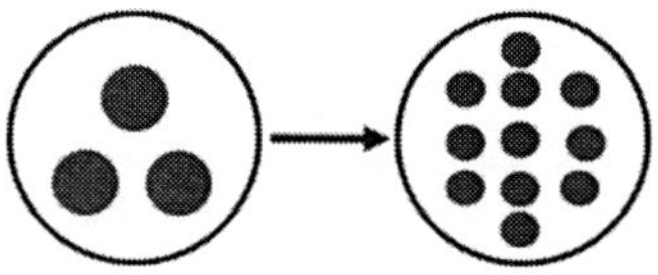

If $\quad n_{d,k} \geq n_{inital}(\frac{d_{inital}}{d_{d,k}})^3$

$d_{d,k2}$ Generated randomly

$$d_{d,k1}^3 = d_{d,k}^3 - d_{d,k2}^3$$

$$n_{d,k} = n_{d,k1} = n_{d,k2}$$

$n_{d,k} < n_{inital}(\frac{d_{inital}}{d_{d,k}})^3$

$d_{d,k1}$ Generated randomly

$$n_{d,k1} = n_{d,k}(\frac{d_{d,k}}{d_{d,k1}})^3$$

- **The model is written as an User-defined function (UDF)**

CFD simulation- Emulsions in STR
Fundaments

Eddy life time:

$$u = \overline{u} + u'$$

- Velocity fluctuation violates the droplet consolidating forces and breaks the droplet.
- Turbulence: an actual eddy makes the velocity fluctuation.
- Eddy life time indicates the turbulent velocity fluctuation along the Lagrangian parcel.

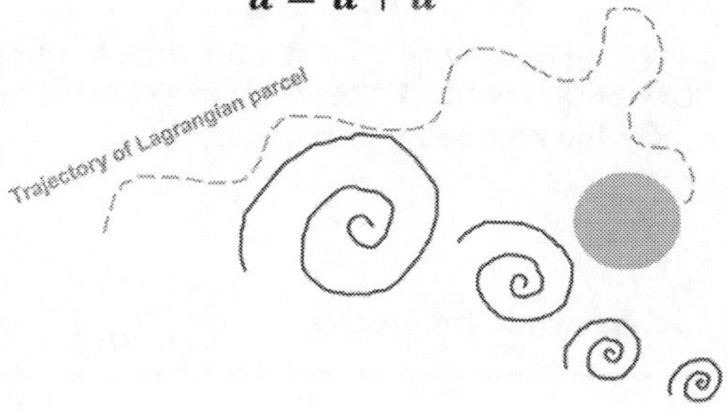

For k - ε turbulence model:

$$\tau_{eddy} = 0.3 \frac{k}{\varepsilon}$$

Therefore, breakup will not take place if the time between two subsequent breakups is less than eddy life time.

- This idea can be aligned with the concept of "Arrival eddy".
- This condition fails if the eddy life time becomes smaller than the numerical time step (Δt).

CFD simulation- Emulsions in STR
Fundaments

Which factors are affecting the droplet size?

Gibbs free energy

$$dG^{\sigma} = -S^{\sigma} dT + A\, d\gamma + \sum n_i d\mu_i$$

- Surface chemistry and physic
- Temperature
- Mass transfer
- Molecular forces (e.g. Covalent, Van der Waals,...)
- Fluid properties such as viscosity, interfacial tension
- Surfactants/additives

What is the main mechanism of droplet breakup?

- Consolidating forces vs disruptive forces*

Our aim is to define a correlation for the local Sauter mean diameter based on the fluid properties and the fluid dynamics

*Hinze, J. O. (1955). Fundamentals of the hydrodynamic mechanism of splitting in dispersion processes. *AIChE Journal*, 1(3), 289-295.

CFD simulation- Emulsions in STR Fundaments

Novel coalescence model

Lagrangian tracers do not have data about parcels in their surrounding, since collisions are not resolved. → how we can estimate collision frequency?!

- **By the mapped volume fractions**

*Coalescence frequency: $\Gamma(d,d') = \zeta(d,d')\lambda(d,d')$

*Collision frequency: $\zeta(d,d') = c_1 \frac{\varepsilon^{1/3}}{1+\phi}(d+d')^2(d^{2/3}+d'^{2/3})^{1/2}$

2 Typo mistake!

*Coalescence efficiency: $\lambda(d,d') = exp(-c_2 \frac{\mu_c \rho_c \varepsilon}{\sigma^2(1+\phi)^3}\left(\frac{d.d'}{d+d'}\right)^4)$

However we used the wrong one!, since c_2 is founded based on it

*Coulaloglou, C. A., & Tavlarides, L. L. (1977). Description of interaction processes in agitated liquid-liquid dispersions. *Chemical Engineering Science, 32*(11), 1289-1297.

CFD simulation- Emulsions in STR Fundaments

- **Actual Lagrangian tracer collides with each coalescence partner (imaginary bins) and the Created droplet will be stored in a new bin based on its diameter class**

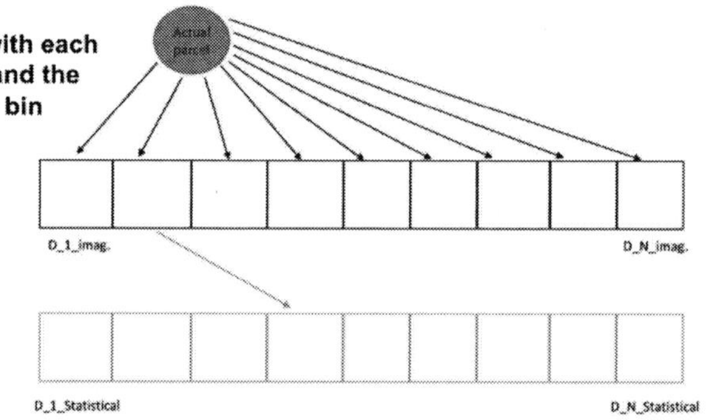

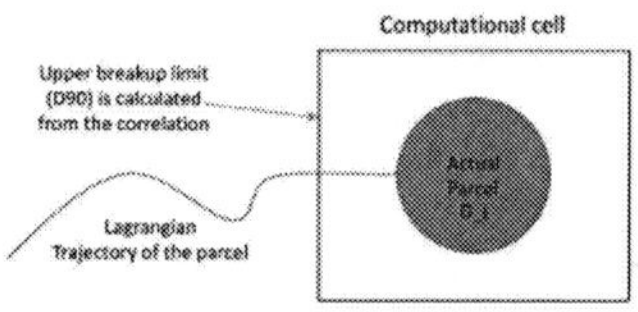

Bus stop model:

- **Saved volume of the created droplet wait for a suitable parcel which has the same dimeter class to be unloaded on that parcel.**
- **This strategy avoids an enormous number of parcels which should be injected in The next time step.**
- **New parcels must be injected if the suitable parcel does not exist in the system.**

CFD simulation- Emulsions in STR
Sub-grid issue

TFM simulation: Do the clusters of dispersed phase form?

2D channel: fine grid - shear driven: Turbulent flow

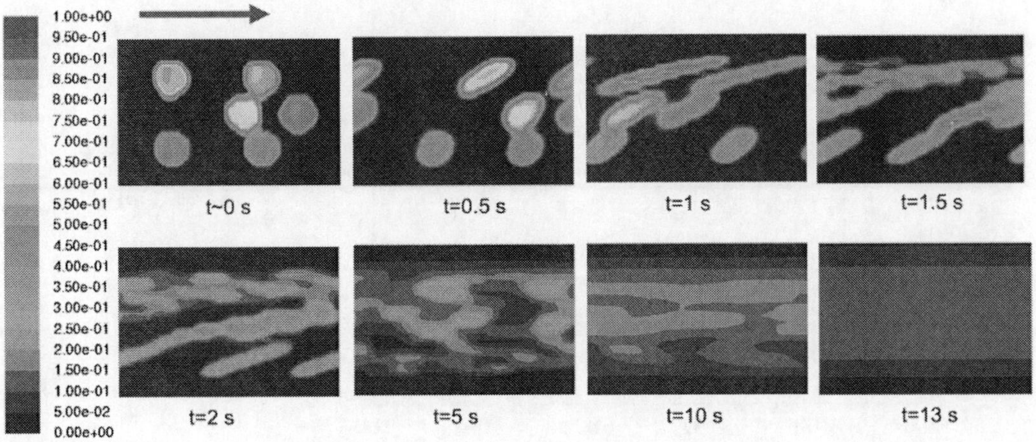

Contour plots of dispersed phase volume fraction

CFD simulation- Emulsions in STR
Results

Impact of the droplet swarm force:

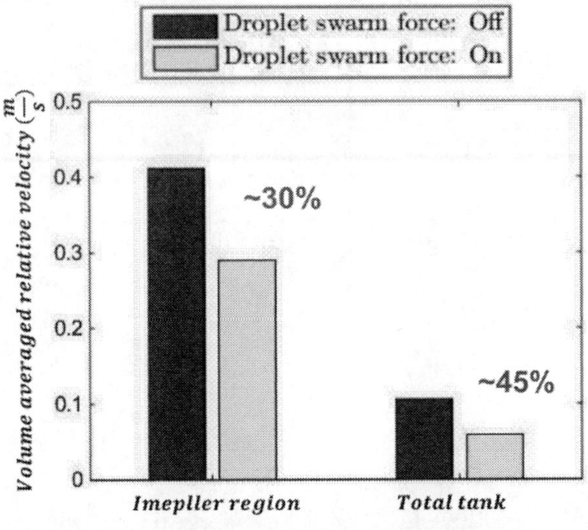

Water-in-Conroe oil at 300RPM

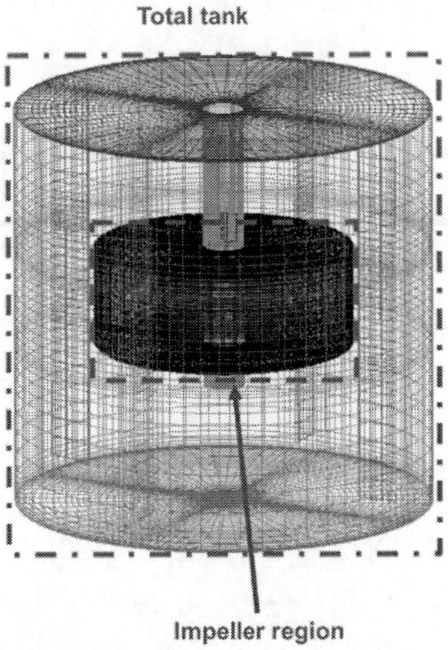

CFD simulation- Emulsions in STR
Results

Eddy life time: Water-in-Conroe Oil

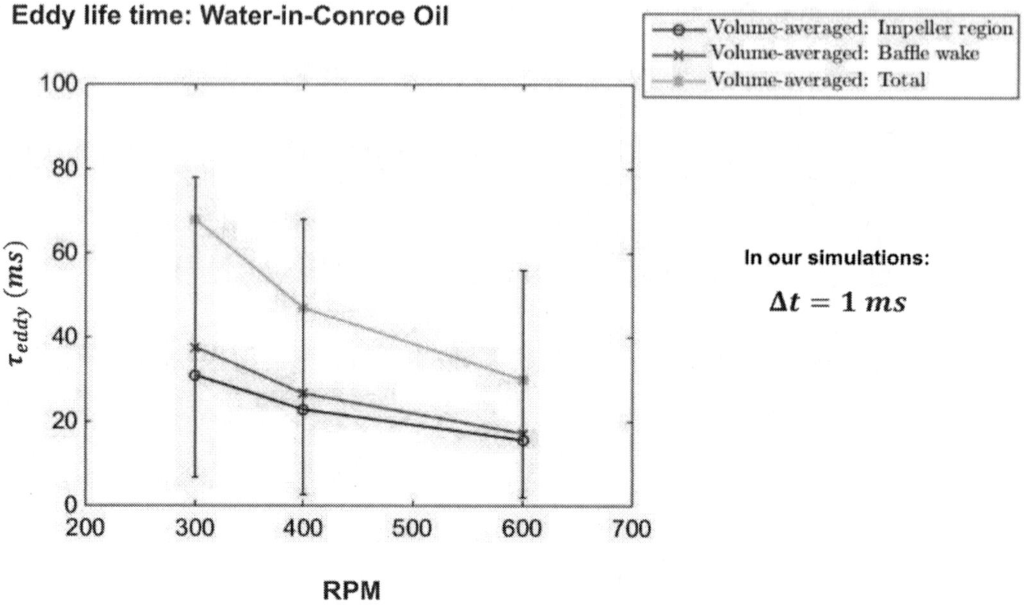

In our simulations:

$$\Delta t = 1\ ms$$

CFD simulation- Emulsions in STR
Results

Breakup limit: Water-in-Conroe Oil

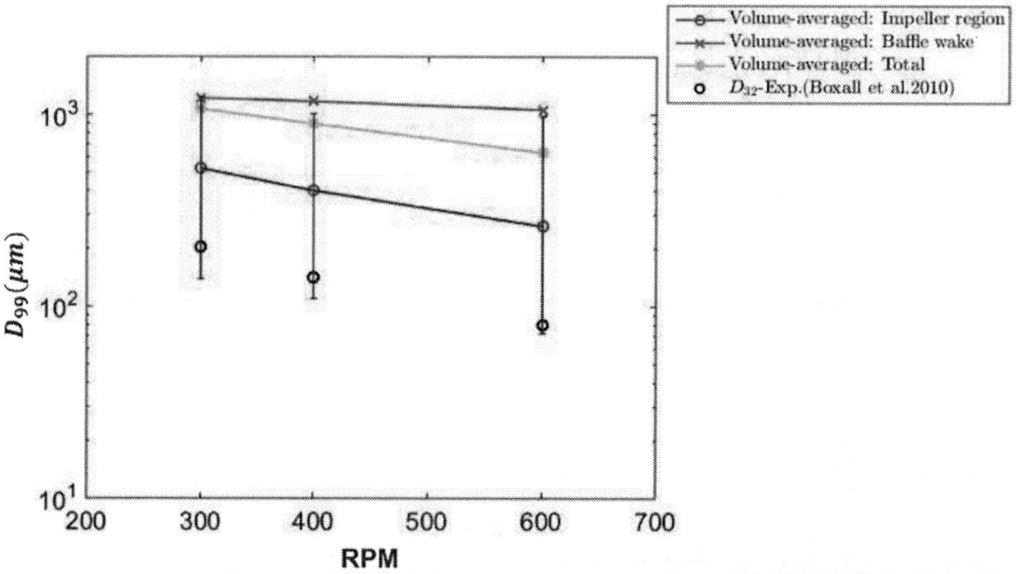

CFD simulation- Emulsions in STR Results

Results of Breakup (coalescence is neglected) : Water-in-Conroe Oil

Initial state

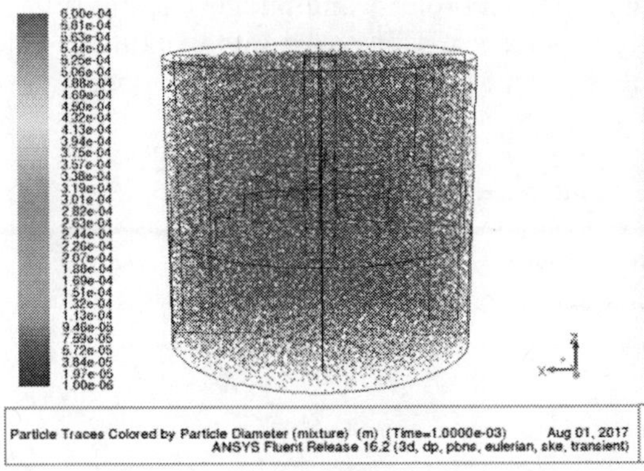

Final state

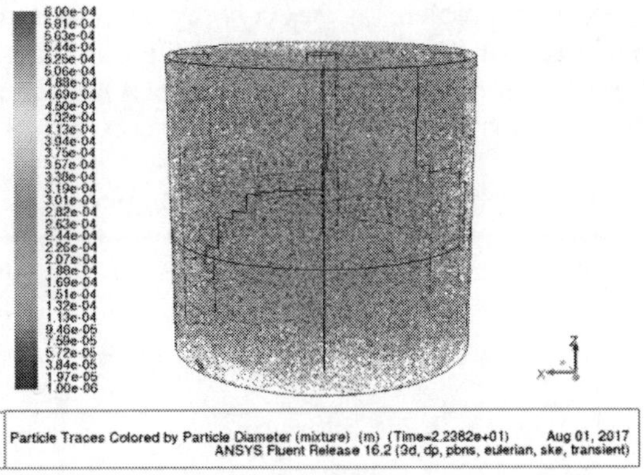

$\approx$ 52,000 parcels ⟶ **$\approx$ 160,000 parcels**

On average $\approx$ $10\sim10^3$ droplets/parcel ⟶ **$\approx$ $10^7\sim10^9$ droplets in system** ⟶ **One step closer to reality**

300 RPM

CFD simulation- Emulsions in STR Results

Results of Coalescence: Water-in-Conroe

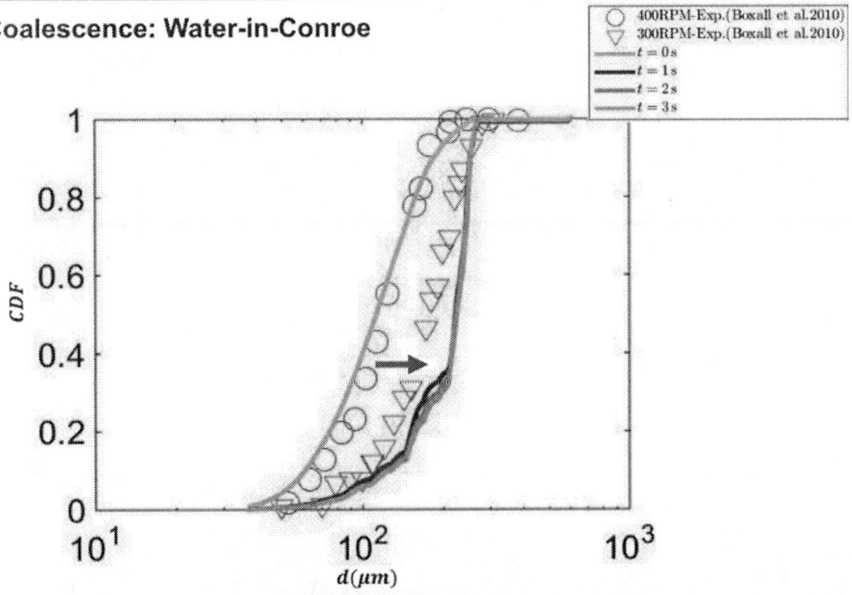

From 400 RPM to 300 RPM

509054 Solve this! Fundamental Approach to Problem Solving in Industrial Processes I (Invited Talks)

Abstract:

Solving complex plant problems is an everyday exercise that affects all projects and budgets, and it is very rarely introduced to students at any level. In fact, by the time young engineers are few years into working in industry, most have lost the engineering skills that they developed in school. We hope to encourage future engineers to retain those skills and learn the importance of plant problem-solving.

503576 Industrial-Scale Modeling of Herschel-Bulkley (Yield Stress) Fluids

Abstract:

The suitability of lattice-Boltzmann-based CFD for modeling Herschel-Bulkley fluid is examined. Under the Herschelâ€“Bulkley model, the strain experienced by a fluid is related to the stress via a yield shear stress, flow consistency, and flow index. Although common in industrial practices, the yield stress-induced discontinuity in the constitutive relationship is difficult to model using time-averaged, RANS-based CFD. The time-accurate nature of the lattice-Boltzmann approach, combined with the high spatial resolution it affords, enables a DNS approach to modeling momentum transport through such systems. Here we assess the industrial suitability of this lattice-Boltzmann DNS approach for modeling Herschelâ€“Bulkley fluids, as applied to draining storage tanks, flow through pipes, and flow inside agitated systems.

499620 Viscometer for on-Site Rheology Measurements:Â Development and Application

Abstract:

It is vitally important to have accurate rheological measurements for fluids when designing equipment for processes requiring mixing of viscous, non-Newtonian fluids. In many cases, it is convenient to send samples to a laboratory and have the measurements made there but there are certain processes in which the properties of the fluid change so rapidly in time that representative measurements can only be made at the site where the process is carried out.

This presentation will discuss the development of a portable viscometer which can be easily transported to a site and used to measure rheology. A case study will be presented demonstrating how the viscometer has been used to investigate problems in a high solids concentration slurry mixing process.

502086 Using Î¼²rheology to Measure Rheological Properties during Repeated Phase Transitions of Hydrogenated Castor Oil

Abstract:

Rheological modifiers are often used as ingredients in home care goods. These materials undergo phase changes during manufacturing and regular use. Rheological and material properties can be altered by repeated phase transitions, and it is important to know the extent to which microstructural changes affect the performance of the end use product. In this work, we investigate a hydrogenated castor oil (HCO) gel, a rod-like colloid (aspect ratio 50-2500) that forms a stable aqueous suspension with the addition of a surfactant and undergoes a phase transition due to an osmotic pressure gradient. Î¼²rheology, the combination of microfluidics and multiple particle tracking microrheology (MPT), is used to measure material properties during transition. A new microfluidic device design enables the exchange of fluids while minimizing shear stress on the sample. In MPT, fluorescent probe particles are embedded in the sample and their Brownian motion is recorded using video microscopy. The particle trajectories are then related to rheological properties using the Generalized Stokes-Einstein Relation. MPT data are compared to phases previously determined by Wilkins et. al., 2009. The phases of rod-like systems are network, where the particles have a sample spanning structure; a bundled phase where rods do not create a network; and a transitional phase where the colloidal particles partially lose scaffold structure and begin forming bundles. MPT results show that a 4wt% HCO gel undergoes a repeatable phase transition between the network and transitional phases due to the osmotic gradient, but never reaches the bundled phase. A 0.125wt% solution of colloidal rods that was prepared using a large amount of shear undergoes repeatable phase transitions but does not form the network structure measured in the 4wt% gel. This is due to the shear added during sample preparation permanently breaking the network structure. This work shows that sample preparation can have irreversible effects on the microstructure of the gels, which can cause changes in end use products made with rheological modifiers.

499275 Model Systems for Mixed Particulate Fermentation Broth

Abstract:

Model systems for mixed particulate fermentation broth

C Bliatsiou, L Böhm, M Kraume

Chair of Chemical and Process Engineering, Technische Universität Berlin, Germany

FH6-1, Fraunhoferstrasse 33-36, 10587 Berlin, Germany. Tel: +493031472791, email: lutz.boehm@tu-berlin.de

The project which is part of the Priority Programme "SPP1934 DiSPBiotech" is concerned with the experimental and numerical characterization of local mechanical stresses occurring in gassed stirred fermenters and their effect on biological agglomerates. The findings thus obtained are practically tested on the fermentation of Aspergillus niger. Furthermore, the results are used for the identification and development of low-shear stirrers.

To create controlled and repeatable conditions in experiments, suitable model systems which represent the biological broth are of high interest. The material properties such as density and rheology but also characteristics of the agglomerates depend on numerous parameters such as biological behavior, the mechanical and operational process settings as well as time. Even for such well known processes as the fermentation of A.niger, material property data can rarely be found. It is known that, depending on the conditions, the broth can have Newtonian (e.g., for pellet growth; Fig.1a) or non-Newtonian characteristics (incl. shear thinning, shear thickening, viscoelastic, e.g., for mycelia growth; Fig.1b) or even changing rheology over time due to increasing biomass concentration in the batch process. The same is valid for the characteristics of the agglomerates with rather large round particles in case of pellets down to smaller, in parts branched, filamentous shapes. This broad variety of conditions in real fermentation broth cannot all be covered by one model system. The requirements that the model solution should fulfill are challenging. It should be able to mimic all the material properties of the real system while covering the requirements of the applied measurement techniques as well. To name one example, e.g., for optical measurements, the continuous phase must be transparent.

Here, material and particle information is collected from literature and generated by own work and together with project partners within the frame of the Priority Programme. This information will be used to find proper model solutions. For a Newtonian rheology of the continuous phase, aqueous solutions with different additives can be used. Furthermore, for high viscous conditions, silicone oil or glycerol are valid options. To generate a non-Newtonian behavior carboxyl cellulose, xanthan gum or polyacrylamid solutions show a shear-thinning and, in parts, additionally viscoelastic behavior while still being transparent.

Covering the task of finding a model system for the dispersed phase as found in real systems with a comparable morphology and morphological behavior during the application of mechanical stress is an even the more complicated task. Suitable options are, e.g., a second immiscible liquid phase (Fig.1a) or blue-clay flocs (Fig.1b). The latter one, especially, is highly promising. Still it must be kept in mind that the dispersed phase can influence the rheology of the whole mixture, as well. To find a guideline for choosing the best model solution representing a particulate fermentation broth, these options must be investigated with according techniques for the determination of material properties, such as rheometers. Here, the accuracy of the measurements is of high importance especially for liquid-solid-mixtures as the solids can settle which changes the properties of the sample over time. As mentioned above, in the frame of this project, the morphology and morphological behavior during the application of mechanical stress, as found in stirred tanks, is of highest interest. This will be investigated in the stirred tank by using an endoscope measurement techniques together with an automated image analysis delivering the particle properties.

Figure 1. Images of a) pellet-like growth of A.niger (upper left) and an according potential model system (liquid-liquid dispersion) and b) of branched filamentous agglomerates of the fungi (upper left) with an according image of blue-clay flocs in a stirred tank

Acknowledgments

Financial support by the Deutsche Forschungsgesellschaft (DFG) in the frame of the Priority Programme SPP1934 DiSPBiotech is gratefully acknowledged. Special acknowledgements go to the working group of Prof. Krull from Technische Universität Braunschweig for the support in the fermentation and imaging of A.niger.

489218 Polycarboxylate Ether Cement Superplasticizers Containing Copolymer Backbones with Improved Rheology Performance

Abstract:

Polycarboxylate ether (PCE) superplasticizers are used in cement to reduce the water level necessary to achieve flow and the desired open time for workability, thus reducing the impact of waterâ€™s deleterious effect on concrete strength. Standard PCEs used in cements are graft copolymers consisting of a polyacrylic acid (PAA) backbone with acrylic ester units containing polyethylene glycol (PEG) side chains. Unfortunately, while PAA-based PCEs show equal or better performance initially, the alkalinity of the cement leads to rapid hydrolysis and loss of slump flow. We have found that using an acrylic acid copolymer backbones to produce PCEs provides excellent flow and open time characteristics. The acrylic acid copolymers are statistical copolymers of acrylic acid and hydrophobic acryloyl morpholine (ACMO). This acrlylic acid and acryloyl morpholine copolymer based PCE has shown extended open time for workability, indicating better stability in cement formulation than PAA based PCE.

490341 Investigation of Blend Time for Fully Turbulent Newtonian Fluids in Stirred Tanks

Abstract:

Blending, the process of combining miscible components into homogenous solutions, is a common unit operation performed by rotating agitators. Industrialists are interested in minimizing the blend time to decrease operating periods and increase throughput while also minimizing capital and operating costs. Therefore, agitator manufacturers are driven to provide efficient and cost effective solutions for blending applications as well as accurate blend time predictions.

A blend time investigation for stirred tanks is being conducted to study the effect of impeller diameter (D = T/2, T/3, and T/5), impeller off-bottom distance (C = T/2, T/3, and T/5), and impeller type (A310 Hydrofoil, Pitched-Blade Turbine, and Rushton Turbine) while considering tank diameter (T = 1.22 m) and mean specific energy dissipation (Îµ = 0.005 and 0.01 W/kg) for fully turbulent Newtonian fluids. The response of conductivity probes to the injection of a saturated NaCl tracer is monitored to experimentally measure blend time forÂ 54 unique configurations. The results of the experimental blend time measurements will be analyzed to determine trends and compared to Grenville's existing well-established blend time correlation.

RANS-SLFM and LES-SLFM numerical simulations of turbulent non-premixed oxy-fuel jet flames using CO2/O2 mixture

Adel Alghamdi, Rodney Fox, Alberto Passalacqua, Dennis Vigil
Iowa State University

Malik Hassanaly, Yihao Tang, Venkat Raman
The University of Michigan

IOWA STATE UNIVERSITY
College of Engineering

Content

- Motivations, Background and Literature
- Objectives and Research Hypothesis
- Transport Equations For RANS and LES
- Steady Laminar Flamelet Model (SFLM)
- Simulation Setup RANS and LES
- Results For B1 Flame From RANS/LES
- The Influence of Scalar Dissipation Rate
- Effects of Inlet Velocity on Flame Structure
- The Influence of Differential Diffusion
- Conclusion

2

IOWA STATE UNIVERSITY

Motivations, Background and Literature

- Oxy-combustion mode prevent NOx formations from going into the atmosphere.
- Oxy-combustion uses CO2 in the oxidizer stream that can helps with Global warming.
- Not widely studied in the academic literature, specifically B flames using SLFM.
- Improve the efficiency of carbon sequestration technology to capture more CO2.

3

IOWA STATE UNIVERSITY

Motivations, Background and Literature

- Nowadays, Efficient oxygen/air separation systems exists, which makes oxy-fuel combustion economically feasible.
- Use the model for more complex chemistry and geometry such as furnaces, boilers, gas turbines etc..
- Improve the efficiency of the oxy-fuel combustion mode in general, that leads to less production of harmful emissions.

4

IOWA STATE UNIVERSITY

Motivations, Background and Literature

Oxy-fuel

Name	mol% H2	Re_f	Jet speed (m/s)	Coflow speed (m/s)	Adiabatic temperature (K)
A-1	55	15000	98.2	0.778	2250
A-2	45	15000	84.5	0.755	2243
A-3	37	15000	75.8	0.739	2236
B-1	55	12000	78.6	0.622	2250
B-2	55	15000	98.2	0.778	2250
B-3	55	18000	117.8	0.933	2250

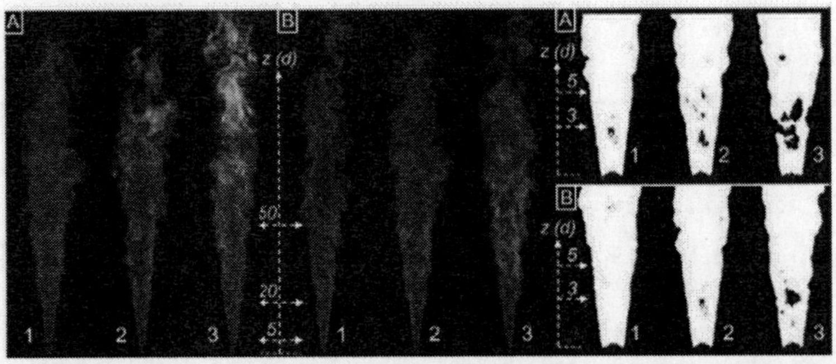

Alexis Sevault a, Matthew Dunnb, Robert S. Barlow b, Mario Ditaranto c, (Combustion and Flames)2011

5

IOWA STATE UNIVERSITY

Motivations, Background and Literature

- Proposed model by (Garmory & Mastorakos 2015) simulating flames A1-A3 using LES/CMC.
  - Capture extinction with reduced chemical mechanism (ARM1).
  - Overall, the model showed good agreement with the conditionally averaged temperature.
  - Inaccurate prediction for O_2 and H_2O mass fraction profiles.
  - Model over-predicted extinction at both locations z/d=3, z/d=5.
  - Successful capture of the physical location of the adiabatic flame temperature

6

IOWA STATE UNIVERSITY

Motivations, Background and Literature

- Laminar flamelet approach first introduced by (Peters 1984).

Flamelet approach can be classified into :

- Steady/Unsteady Laminar Flamelet Models (SLFM).
- Flamelet Progress Variable model (FPV) by (Pierce&Moin 2003)

- Many established CFD codes have implemented most of flamelet versions. (OpenFoam, Fluent, CFX, Star CCM+)

7

IOWA STATE UNIVERSITY

Objectives and Research Hypothesis

Objectives:

- Use SLFM with RANS/LES to simulate different flame configurations (oxy-fuel flames)

- Predict flame structure, using GRI3.0 to capture the temperature and species distributions.

- Validate results from RANS/SLFM and LES/SLFM against experimental data published in the academic literature.

8

IOWA STATE UNIVERSITY

Objectives and Research Hypothesis

- Investigation of the effects of scalar dissipation rate on the flame structure.
- Test the effect of different jet velocity on the global stability of the flames.
- Study the effect differential diffusion on oxy-fuel flames.

Research Hypothesis

- LES/SLFM should produce accurate predictions of the flame structure and pollutant concentrations compare to RANS/SLFM.
- LES/SLFM should be more sensitive to the change in the inlet conditions, that affect the flame structure.

Transport Equations For RANS

- Mass

$$\frac{\partial}{\partial t}(\bar{\rho})+\frac{\partial}{\partial x_j}\left(\bar{\rho}\tilde{\vec{v}}\right)=0$$

- Momentum

$$\frac{\partial}{\partial t}\left(\bar{\rho}\tilde{\vec{v}}\right)+\frac{\partial}{\partial x_j}\left(\bar{\rho}\vec{v}\vec{v}\right)=-\frac{\partial}{\partial x_j}(\bar{p})+\frac{\partial}{\partial x_j}(\bar{\sigma})-\frac{\partial}{\partial x_j}\left(\bar{\rho}v''v''\right)+(\bar{\rho}\vec{g})$$

- Mixture Fraction

$$\frac{\partial\left(\bar{\rho}\tilde{Z}\right)}{\partial t}+\frac{\partial}{\partial x_j}\left(\bar{\rho}\vec{v}Z\right)=\frac{\partial}{\partial x_j}\left(\frac{\mu_l+\mu_t}{\sigma_t}\frac{\partial}{\partial x_j}\left(\tilde{Z}\right)\right)$$

- Mixture Fraction Variance

$$\frac{\partial}{\partial t}\left(\bar{\rho}Z''^2\right)+\nabla\cdot\left(\bar{\rho}\vec{v}Z''^2\right)=\frac{\partial}{\partial x_j}\left[\left(\frac{\mu_l+\mu_t}{\sigma_t}\frac{\partial}{\partial x_j}\left(Z''^2\right)\right)\right]+C_g\mu_t\left(\nabla\bar{Z}\right)^2-\left(\bar{\rho}\tilde{\chi}\right)$$

Transport Equations For RANS

- Kinetic Energy

$$\frac{\partial}{\partial t}\left(\bar{\rho}\tilde{k}\right)+\frac{\partial}{\partial x_j}\left(\bar{\rho}\tilde{\tilde{v}}\tilde{k}\right)=\frac{\partial}{\partial x_j}\left[\left(\mu+\frac{\mu_t}{\sigma_k}\right)\nabla\tilde{k}\right]+P_k-\left(\bar{\rho}\tilde{\varepsilon}\right)$$

- Dissipation Rate

$$\frac{\partial}{\partial t}\left(\bar{\rho}\tilde{\varepsilon}\right)+\frac{\partial}{\partial x_j}\left(\bar{\rho}\tilde{\tilde{v}}\tilde{\varepsilon}\right)=\frac{\partial}{\partial x_j}\left[\left(\mu+\frac{\mu_t}{\sigma_\varepsilon}\right)\frac{\partial}{\partial x_j}(\tilde{\varepsilon})\right]+\left(c_{\varepsilon 1}\frac{\tilde{\varepsilon}}{\tilde{k}}P_k\right)-\left(c_{\varepsilon 2}\bar{\rho}\frac{\tilde{\varepsilon}^2}{\tilde{k}}\right)$$

- Turbulent Eddy Viscosity $\mu_t = C_\mu \bar{\rho}\frac{\tilde{k}^2}{\tilde{\varepsilon}}$

- Scalar Dissipation Rate $\tilde{\chi} = C_\chi \frac{\varepsilon}{k} Z''^2$

11

IOWA STATE UNIVERSITY

Transport Equations For LES

- Mass $\frac{\partial}{\partial t}(\bar{\rho})+\frac{\partial}{\partial x_j}\left(\bar{\rho}\tilde{u}_j\right)=0$

- Momentum $\frac{\partial}{\partial t}(\bar{\rho}\tilde{u}_i)+\frac{\partial}{\partial x_j}\left(\bar{\rho}\tilde{u}_i\tilde{u}_j\right)=-\frac{\partial}{\partial x_j}(\tilde{p})+\frac{\partial}{\partial x_j}\left(\sigma_{ij}\right)+\frac{\partial}{\partial x_j}\left(\tau_{ij}{}^{SGS}\right)$

- Sub-grid Stress

$$\sigma_{ij}=2\bar{\rho}\upsilon\left[\tilde{S}_{ij}-\frac{1}{3}\delta_{ij}\frac{\partial}{\partial x_k}(\tilde{u}_k)\right],\tilde{S}_{ij}=\frac{1}{2}\left[\frac{\partial}{\partial x_j}(\tilde{u}_i)+\frac{\partial}{\partial x_j}(\tilde{u}_j)\right]$$

$$\tau_{ij}{}^{SGS}=\bar{\rho}\tilde{u}_i\tilde{u}_j-\bar{\rho}u_iu_j=2\mu_tS_{ij}$$

- Mixture Fraction

$$\frac{\partial}{\partial t}\left(\bar{\rho}\tilde{Z}\right)+\frac{\partial}{\partial x_j}\left(\bar{\rho}\tilde{u}_j\tilde{Z}\right)=\frac{\partial}{\partial x_j}\left(\bar{\rho}D\frac{\partial}{\partial x_j}(\tilde{Z})\right)+\frac{\partial}{\partial x_j}\left(M^{SGS}\right)$$

12

IOWA STATE UNIVERSITY

Transport Equations For LES

- Scalar fluxes sub-filter term

$$M^{SGS} = \bar{\rho} u_i Z_j - \bar{\rho} \tilde{u}_i \tilde{Z} = \frac{\mu_t}{\sigma_t} \frac{\partial Z}{x_j}$$

- Closure Models

Dynamic Averaged Smagorinsky subgrid model was used for the unresolved transport terms

$$\mu_t = C_s \bar{\rho} \Delta^2 \hat{S}$$

$$\chi = C_\chi D \left| \nabla Z \right|^2$$

$$Z''^2 = C_V \Delta^2 \left| \nabla \tilde{Z} \right|^2$$

13

IOWA STATE UNIVERSITY

Steady Laminar Flamelet Model (SFLM)

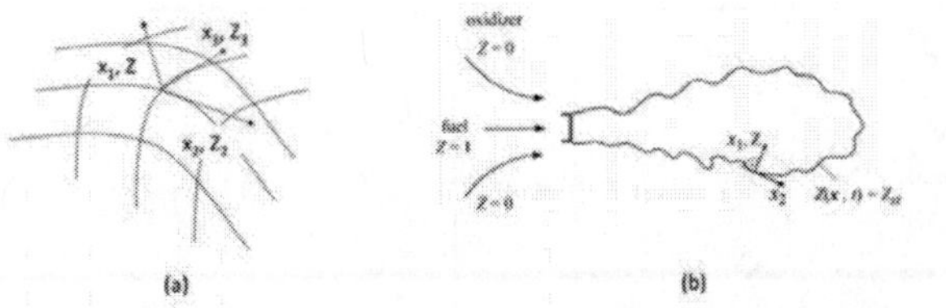

(a) co-ordinate transformation of mixture fraction
(b) surface of stoichiometric mixture fraction in turbulent jet flame (Peters 1984a)

14

IOWA STATE UNIVERSITY

Steady Laminar Flamelet Model (SFLM)

The governing equations

- Temperature

$$\frac{1}{2}\rho\chi\frac{\partial^2 T}{\partial Z^2} - \frac{1}{c_p}\sum_{i=1}^{n} h_i\omega_i + \frac{1}{2c_p}\rho\chi\left(\frac{\partial c_p}{\partial Z} + \sum_{i=1}^{n} c_{p,i}\frac{\partial Y_i}{\partial Z}\right)\frac{\partial T}{\partial Z} = 0$$

- Species Mass Fraction

$$\frac{1}{2}\rho\chi\frac{\partial^2 Y_i}{\partial Z^2} + \omega_i = 0$$

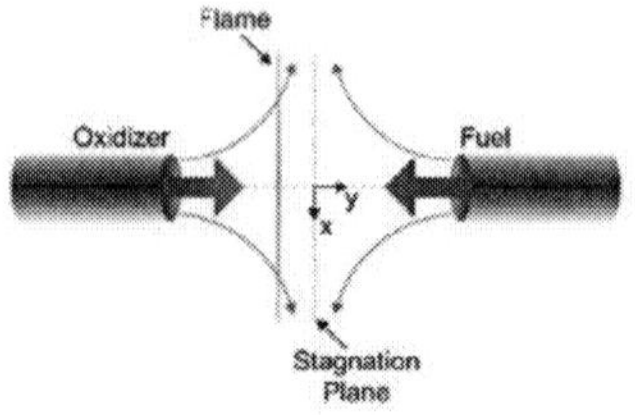

Müller H, Ferraro F, Pfitzner M. Implementation of a Steady Laminar Flamelet Model for non-premixed combustion in LES and RANS simulations. 8th Int OpenFOAM Work 2013:11-2

Steady Laminar Flamelet Model (SFLM)

Modelling flowchart using SLFM with RANS and LES context

Flamelets transport equations:

$$\frac{1}{2}\rho\chi\frac{\partial^2 Y_i}{\partial Z^2} + \omega_i = 0$$

$$\frac{1}{2}\rho\chi\frac{\partial^2 T}{\partial Z^2} - \frac{1}{c_p}\sum_{i=1}^{n} h_i\omega_i + \frac{1}{2c_p}\rho\chi\left(\frac{\partial c_p}{\partial Z} + \sum_{i=1}^{n} c_{p,i}\frac{\partial Y_i}{\partial Z}\right)\frac{\partial T}{\partial Z}$$

→

Solving the flamelets equations

$$\Phi(Z,\chi)$$

↓

Use the PDF statistical approach:

$$\iint \Phi(Z,\chi)P(Z,\chi)\,dZ\,d\chi$$

Tabulate chemistry in terms of its mean

$$\tilde{\phi}(\tilde{Z},\tilde{Z}''^2,\tilde{\chi})$$

←

Pressure based solvers RANS/LES

$$\tilde{\phi}(\tilde{Z},\tilde{Z}''^2,\tilde{\chi}) \rightarrow (\tilde{Z},\tilde{Z}''^2,\tilde{\chi})$$

Steady Laminar Flamelet Model (SFLM)

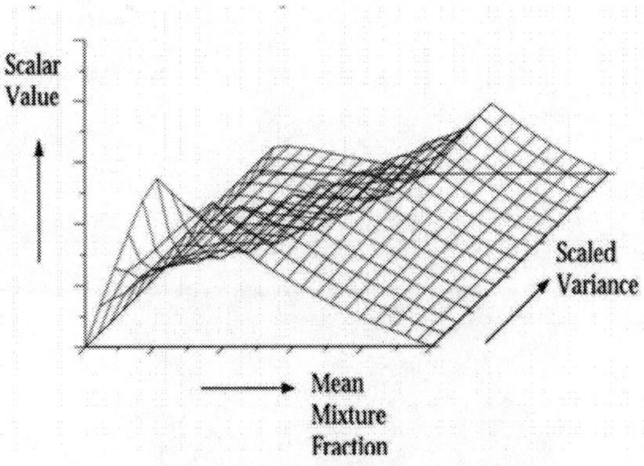

Visual Representation of a Look-Up Table for the Scalar φi as a Function of f and f02 in Adiabatic Single-Mixture-Fraction Systems (fluent manual)

Mean temperature profile for methane/air flame.

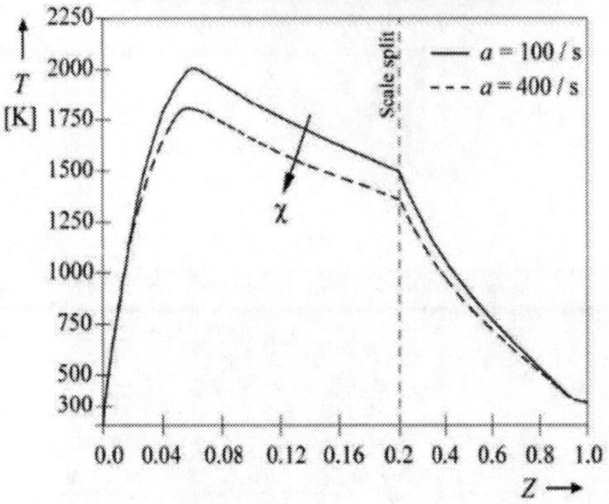

Peters, N. 1992. "Lecture 8 Laminar Diffusion Flames : Diffusion Flamelet Theory."

17

IOWA STATE UNIVERSITY

Simulation Setup (B-1)

Computational domain and simulation conditions:

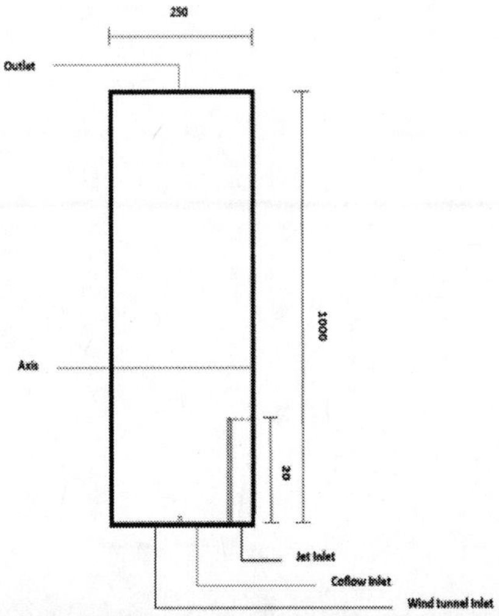

Parameter	Measurements
Fuel jet diameter [mm]	5
Coflow diameter [mm]	96.5
Fuel composition H_2/CH_4 [Mol.%]	55/45
Fuel jet mean velocity [m/s]	78.5
Coflow mean velocity [m/s]	0.622
Oxidizer composition CO2/O2 [Mol%]	68/32
Fuel Temperature[K]	294
Oxidizer Temperature [K]	294
Mesh Number	160K

18

IOWA STATE UNIVERSITY

LES Simulation Setup

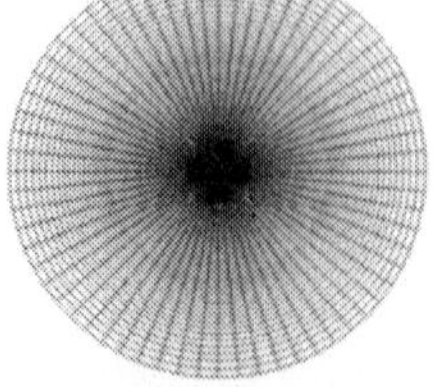

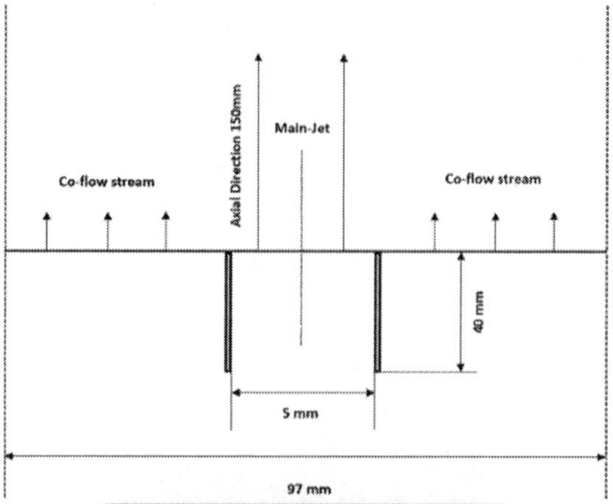

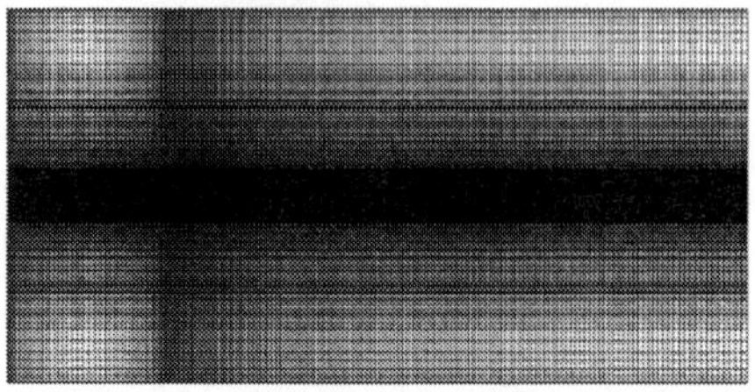

Mesh 0f 2.2M

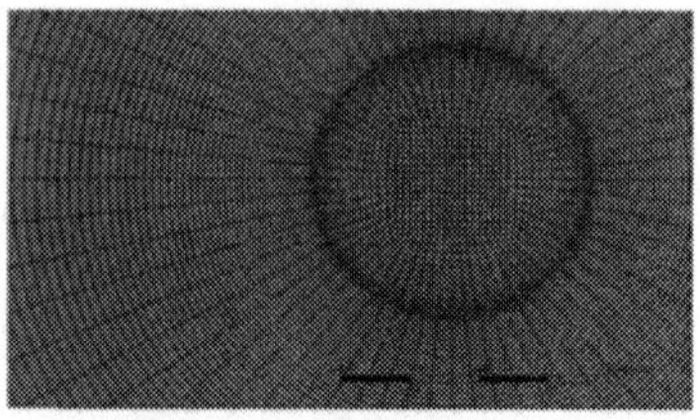

Yplus less than 3

19

(B1) Flame Results From RANS/LES

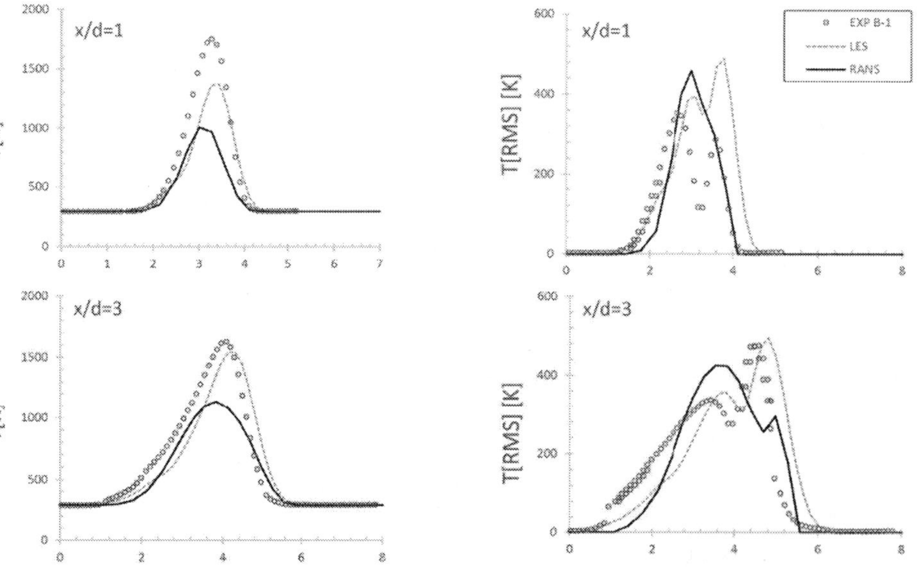

20

(B1) Flame Results From RANS/LES

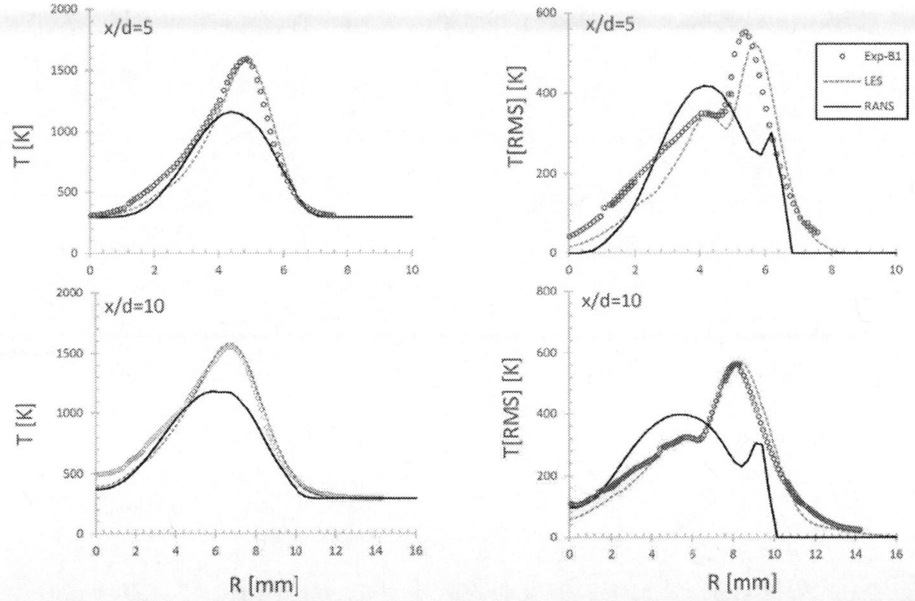

21

IOWA STATE UNIVERSITY

(B1) Flame Results From RANS/LES

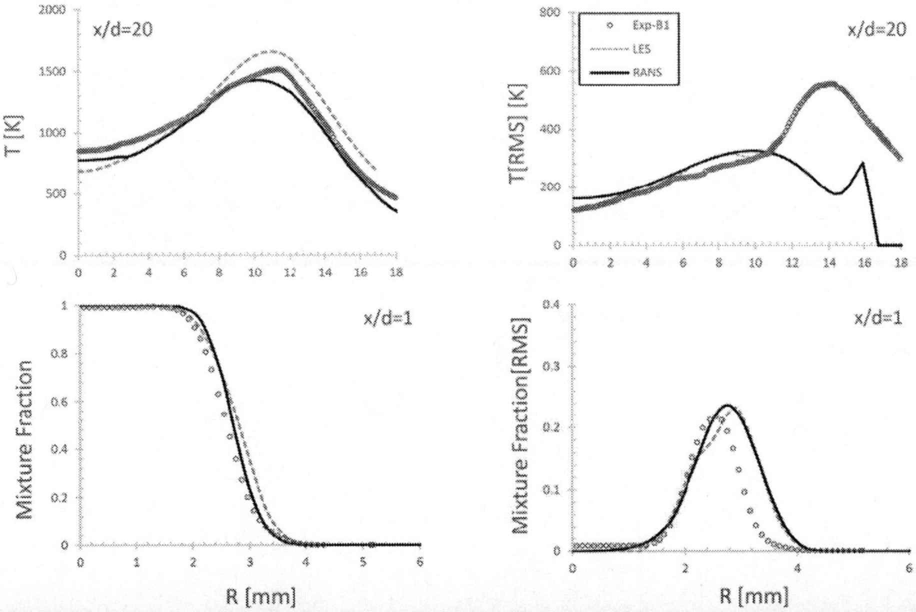

22

IOWA STATE UNIVERSITY

(B1) Flame Results From RANS/LES

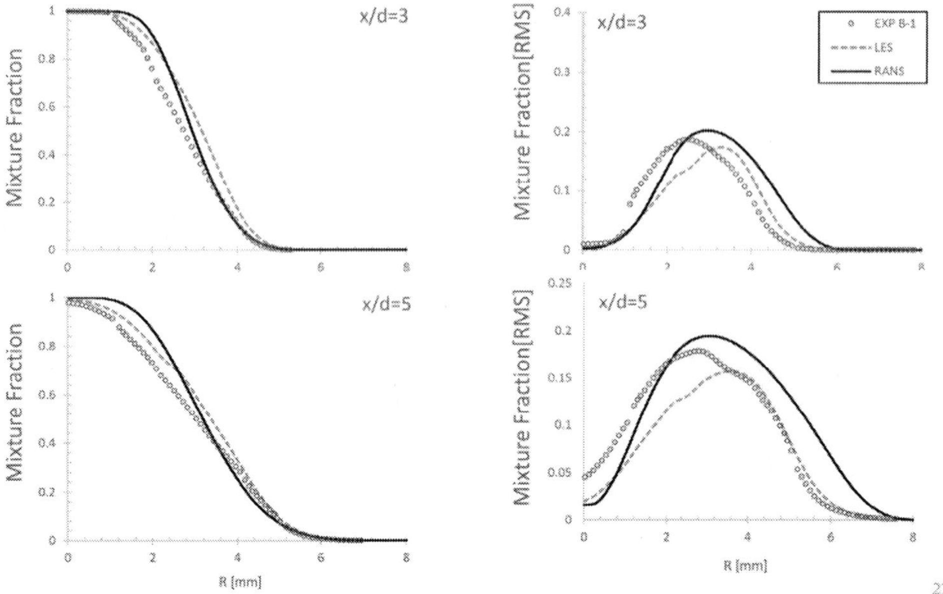

23

IOWA STATE UNIVERSITY

(B1) Flame Results From RANS/LES

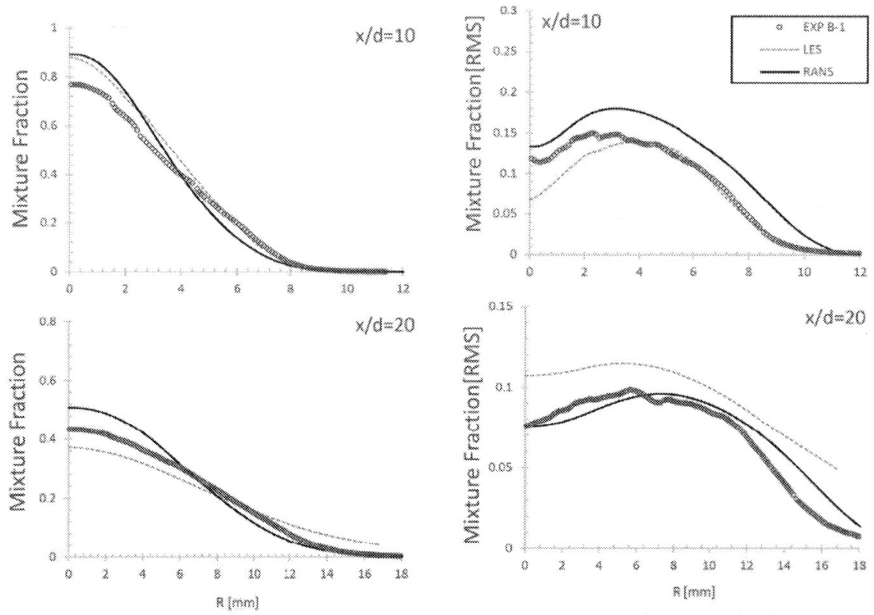

24

IOWA STATE UNIVERSITY

The Influence of Scalar Dissipation Rate

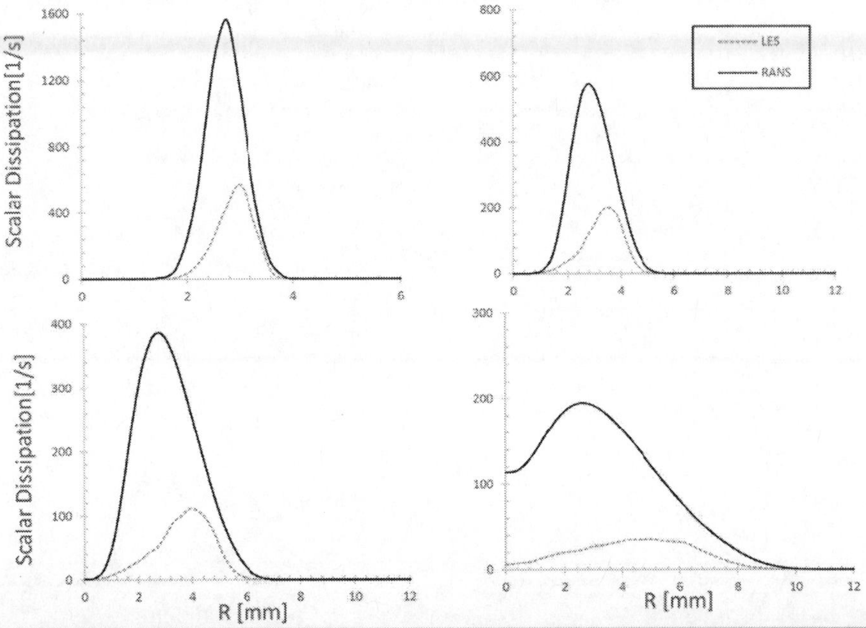

25

IOWA STATE UNIVERSITY

(B1) Flame Results From RANS/LES

Temperature **Mixture Fraction**

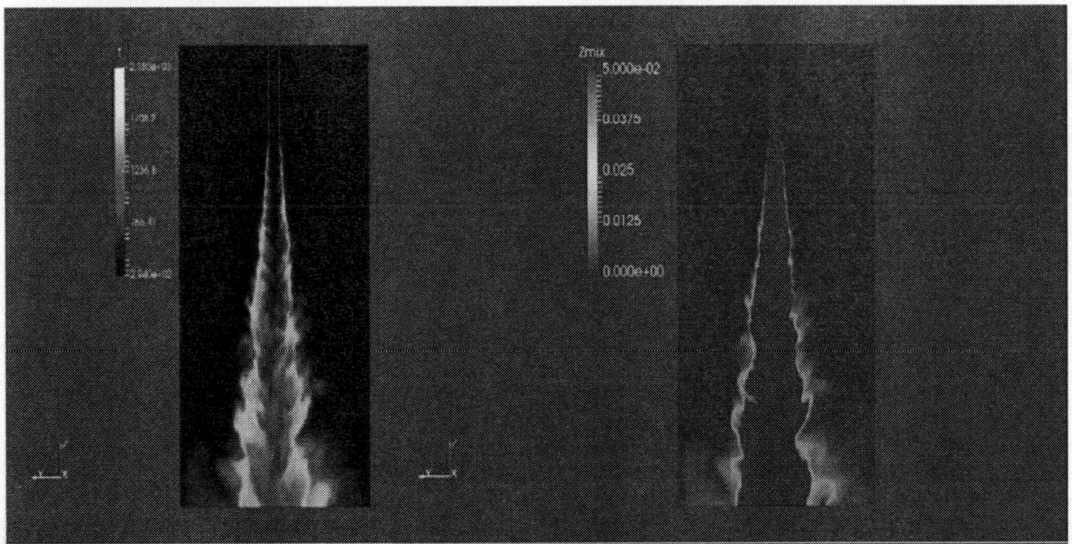

IOWA STATE UNIVERSITY

(B1) Flame Results From RANS/LES

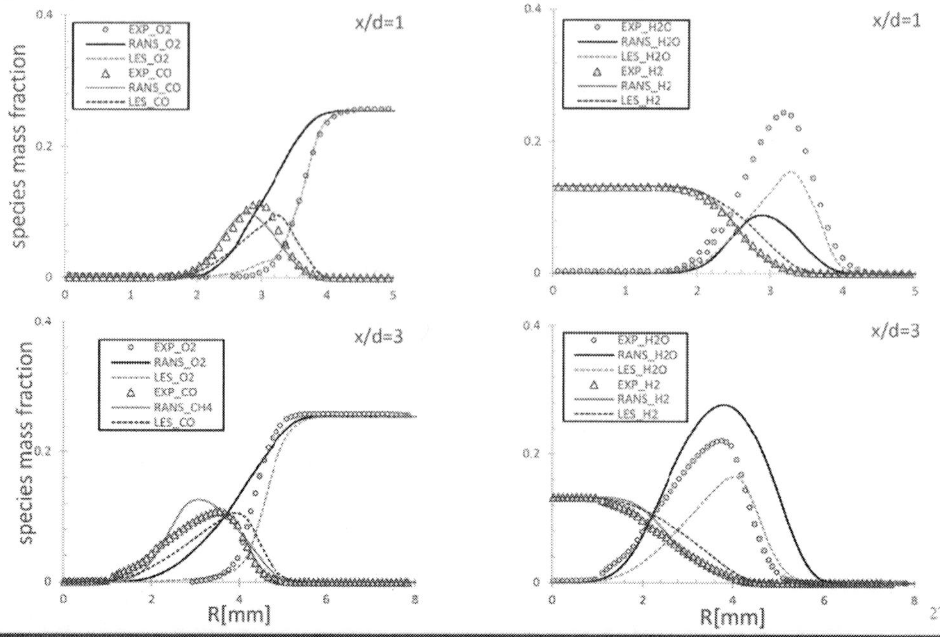

27

IOWA STATE UNIVERSITY

(B1) Flame Results From RANS/LES

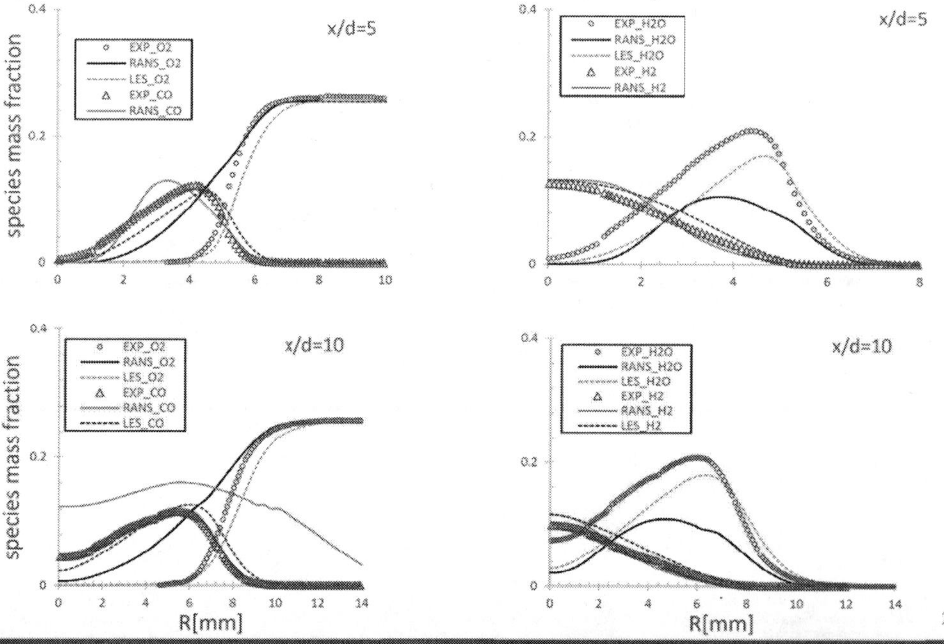

28

IOWA STATE UNIVERSITY

(B1) Flame Results From RANS/LES

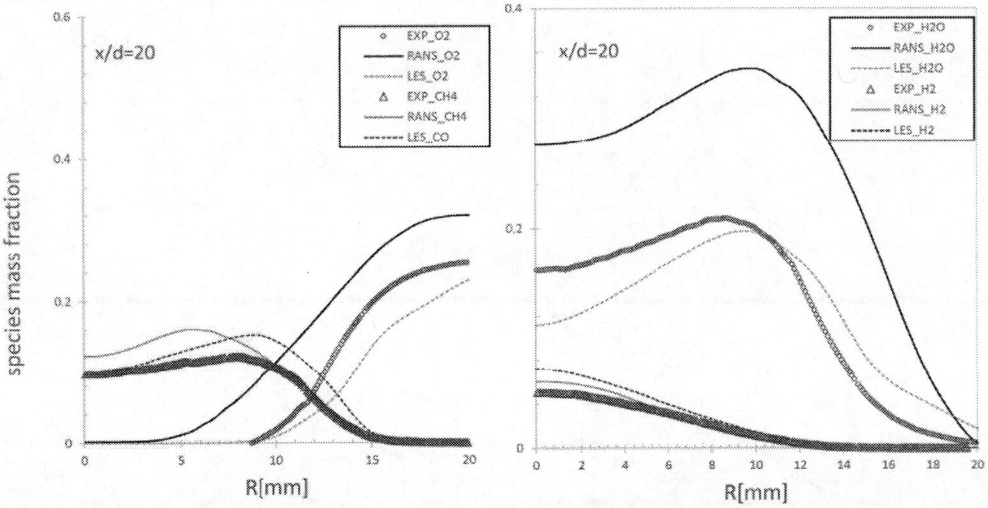

29

IOWA STATE UNIVERSITY

The Influence of Differential Diffusion

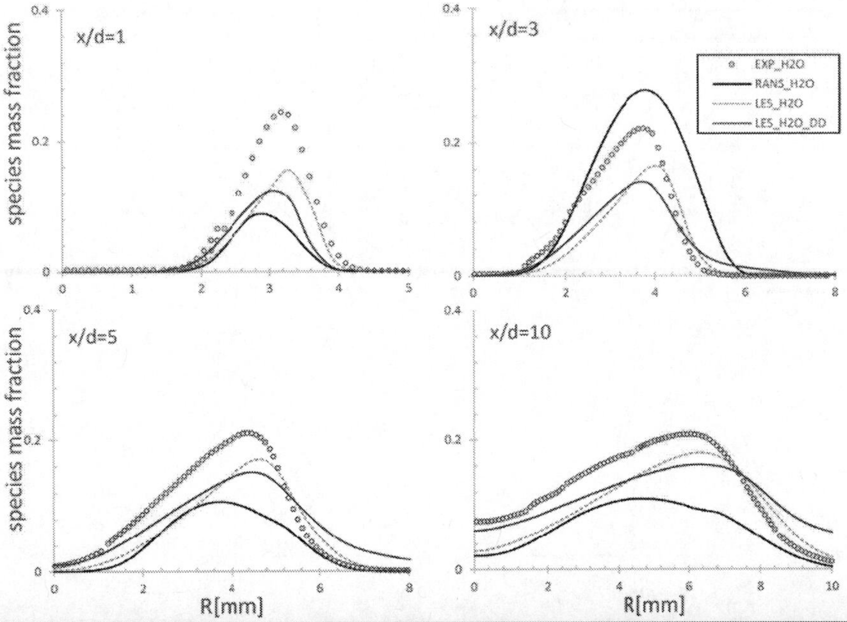

30

IOWA STATE UNIVERSITY

The Influence of Differential Diffusion

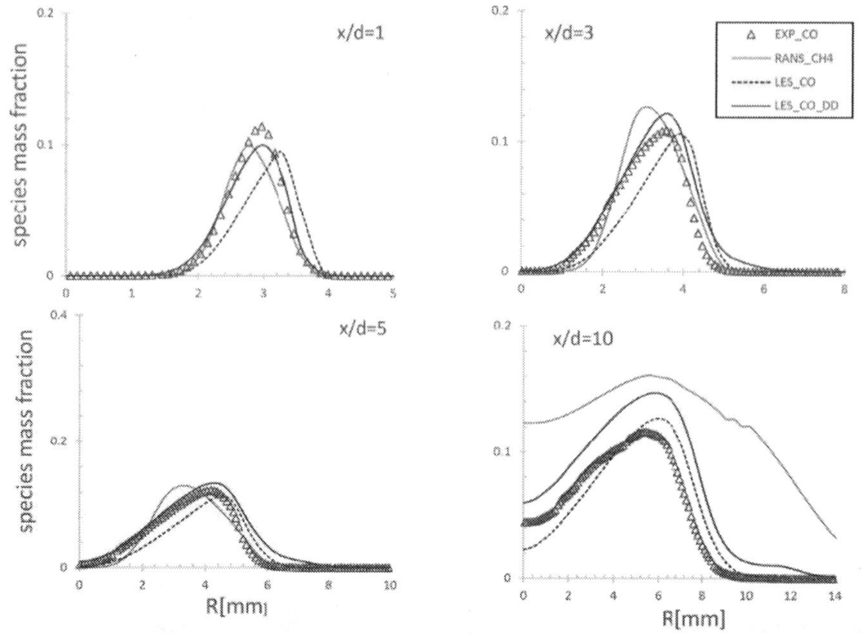

31

IOWA STATE UNIVERSITY

Effects of Inlet Velocity on Flame Structure

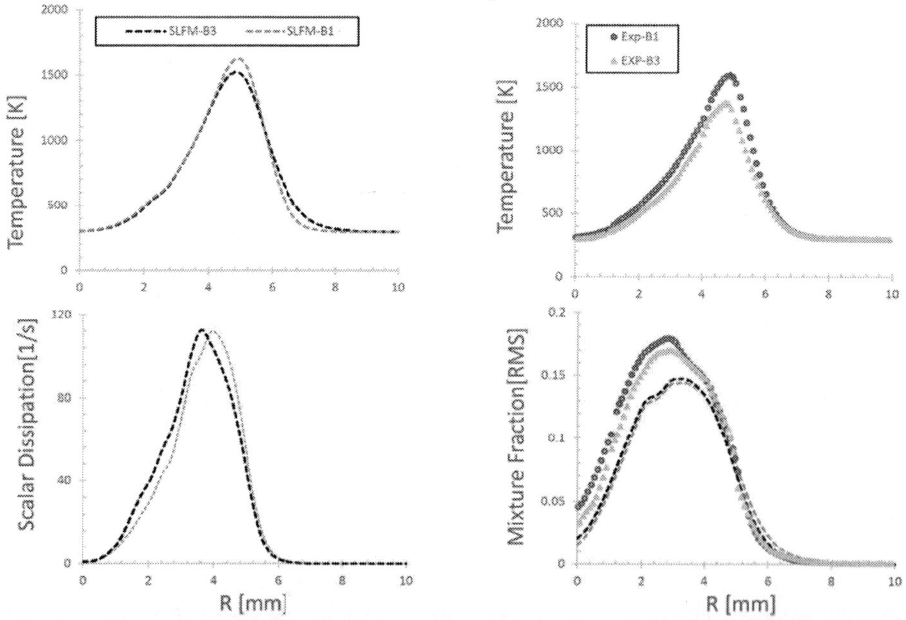

32

IOWA STATE UNIVERSITY

Conclusion

RANS/SLFM

B-1 flame:

- Under-predicted temperature near the nozzle at all locations.
- Reasonable radial profile at the location x/d=20 with the Schmidt number set to 1.6.

33

IOWA STATE UNIVERSITY

Conclusion

LES/SLFM

- LES/SLFM outperform RANS/SLFM in the prediction of B-1 flame scalars.
- The inaccuracy of RANS/SLFM results was due to the overprediction scalar dissipation rate.
- Downstream B-1 flame, all profiles predictions improved since the effect of scalar dissipation rate start to vanish.
- LES/SLFM offer more accurate representation of mixing and no calibration are required for the model constants.
- LES/SLFM helped in calibrating RANS/SLFM model parameters.

34

IOWA STATE UNIVERSITY

Conclusion

- The chemistry of B-1 flame is not well presented by the GRI 3.0 mechanism; it seems that CO2 altered some of the reactions pathways (P. Sabia et.al).

B-3 flame:

- The adiabatic flame temperature was dropped by about 240K at the location of x/d=5, LES/SLFM prediction of the temperature drop was only 100K at the same location.
- SLFM main formulations that do not account for extinction and re-ignition of B-3 flame.

35

IOWA STATE UNIVERSITY

ACKNOWLEDGMENTS

- The financial support for this work was provided by Saudi Arabia Basic Industries Corporation (SABIC).
- The research calculations were conducted using Iowa State University (ISU) HPC resources.
  - Dr. Fox and *Committee members.*
  - Dr. Raman group at University of Michigan (UMICH) .
  - *Graduate students and postdoctoral fellows(ISU).*
  - *SABIC management .*

- Dr. Pannala Sreekanth and David West for initiating this project, and for their valuable feedback and suggestions during the course of this project.

36

IOWA STATE UNIVERSITY

Thank You

37

IOWA STATE UNIVERSITY

(B1) Flame Results From RANS/SLFM

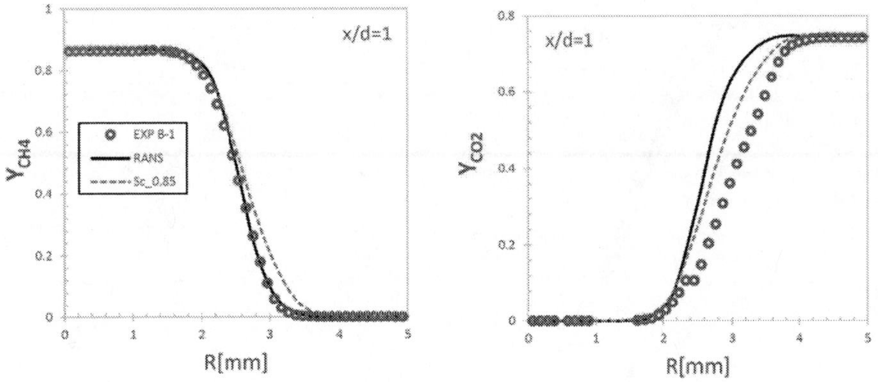

38

IOWA STATE UNIVERSITY

(B1) Flame Results From RANS/SLFM

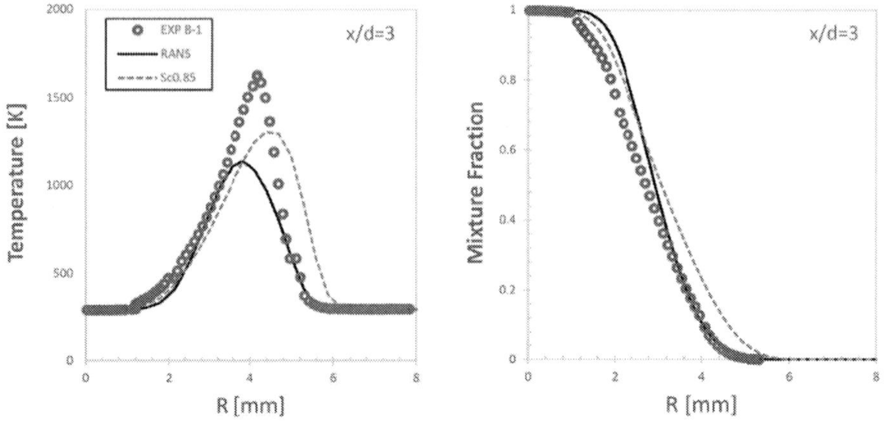

39

IOWA STATE UNIVERSITY

(B1) Flame Results From RANS/SLFM

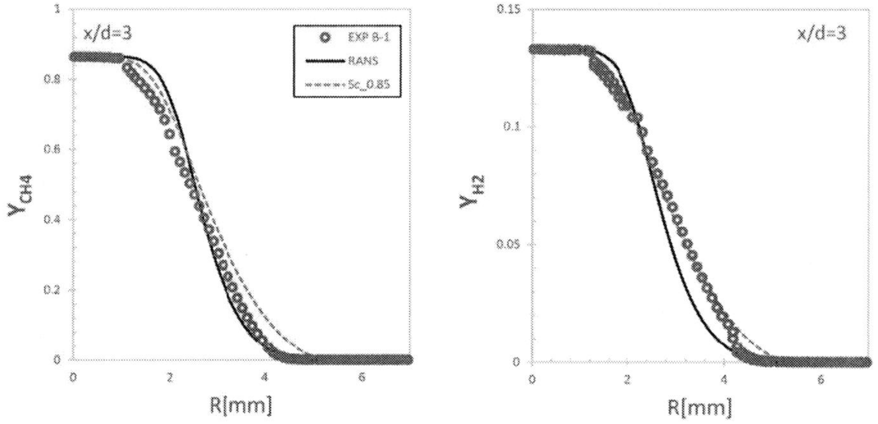

40

IOWA STATE UNIVERSITY

(B1) Flame Results From RANS/SLFM

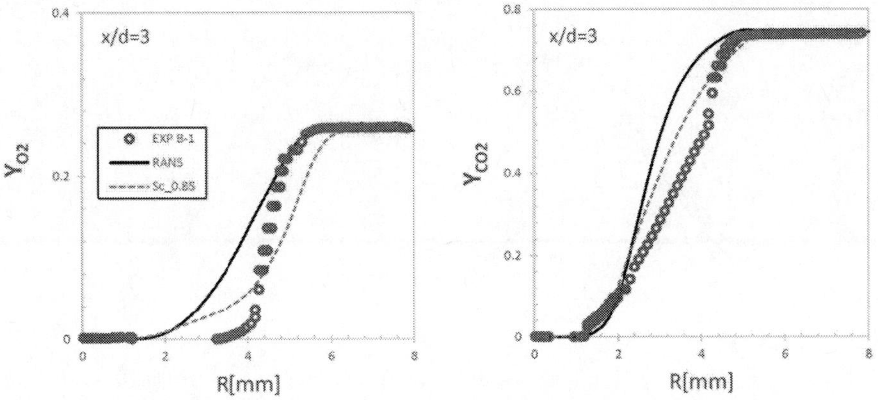

41

IOWA STATE UNIVERSITY

(B1) Flame Results From RANS/SLFM

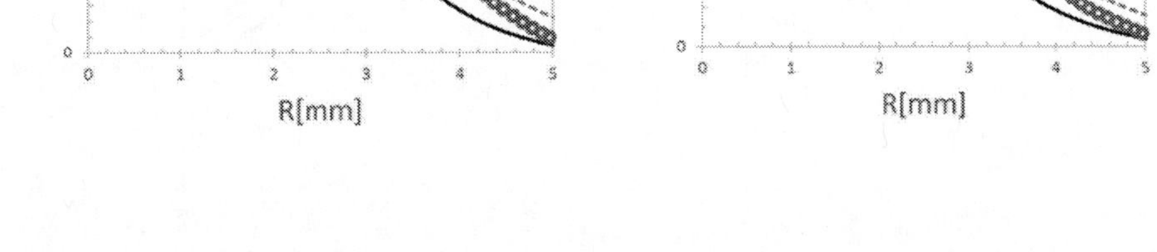

42

IOWA STATE UNIVERSITY

(B1) Flame Results From RANS/SLFM

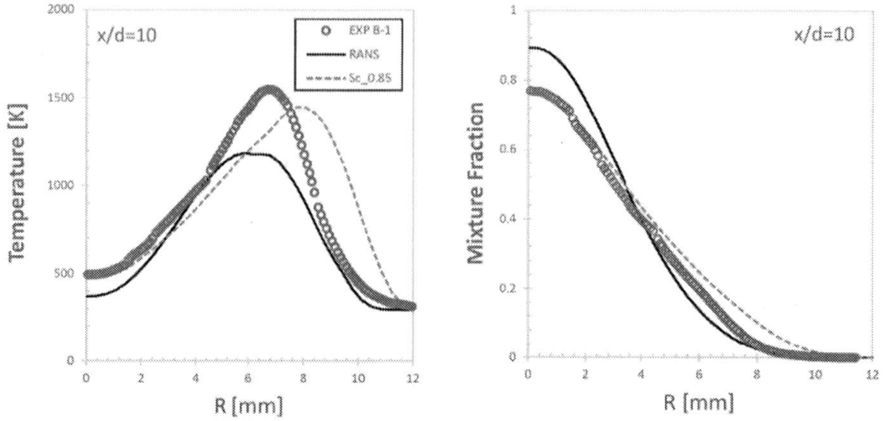

43

IOWA STATE UNIVERSITY

(B1) Flame Results From RANS/SLFM

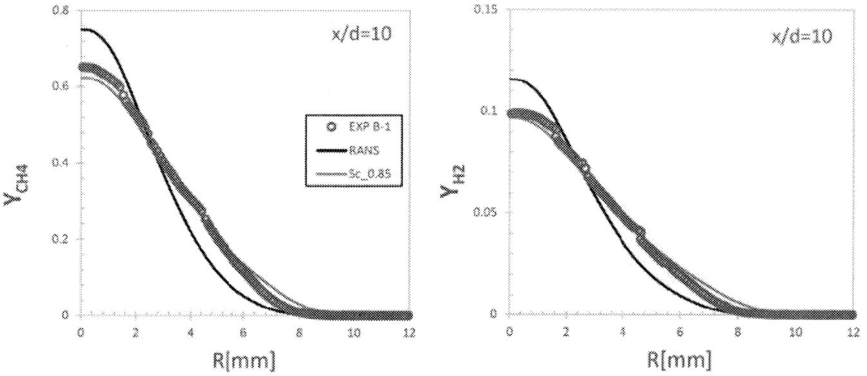

44

IOWA STATE UNIVERSITY

(B1) Flame Results From RANS/SLFM

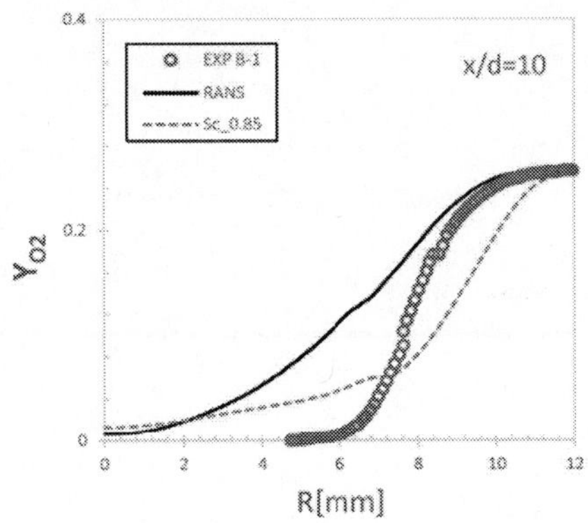

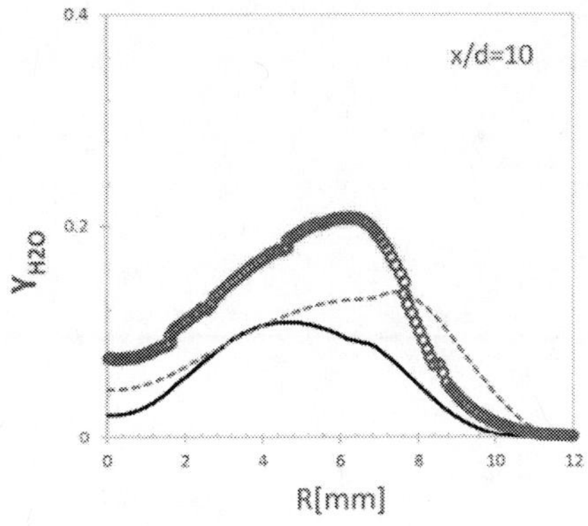

45

IOWA STATE UNIVERSITY

The Influence of Scalar Dissipation Rate

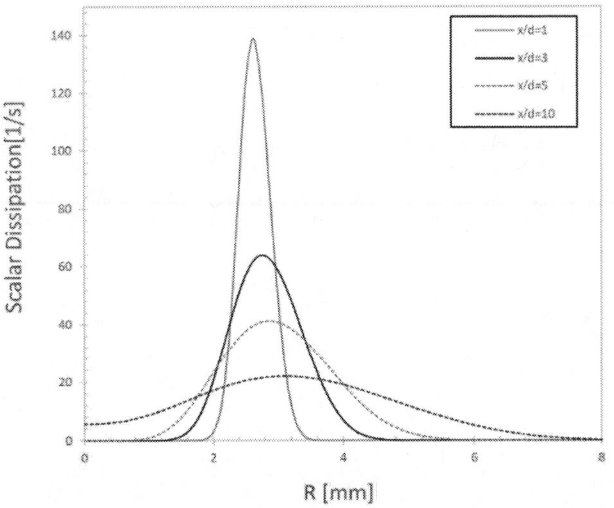

46

IOWA STATE UNIVERSITY

The Influence of Scalar Dissipation Rate

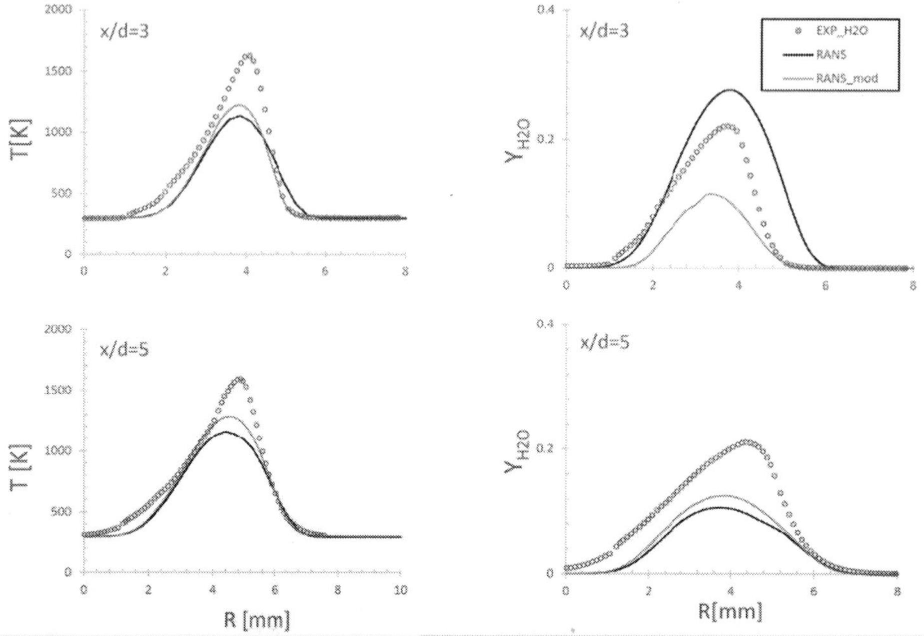

47

IOWA STATE UNIVERSITY

The Influence of Scalar Dissipation Rate

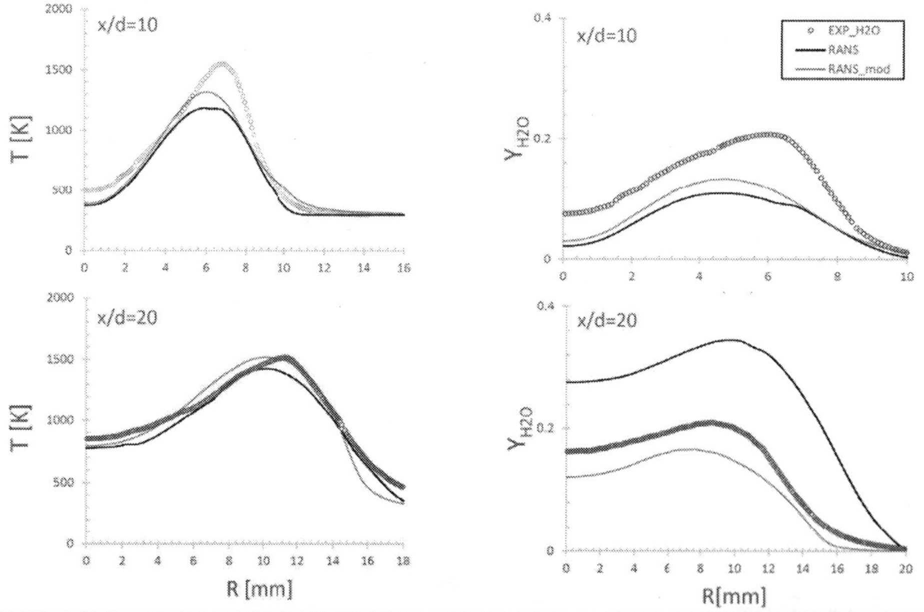

48

IOWA STATE UNIVERSITY

Concluding Remarks

RANS/SLFM

B-1 flame:

- The over-prediction of scalar dissipation rate, affected the temperature and species mass fraction average radial profile.
- Scalar dissipation rate in RANS was over-predicted because of the inaccuracy of the turbulence model.
- Downstream of both flames seems to be affected more by chemistry and not by the flow calculations.

49

IOWA STATE UNIVERSITY

Concluding Remarks

LES/SLFM

Sandia D flame:

- LES/SLFM outperformed RANS/SLFM in the predictions of NO and CO of Sandia D flame.
- Not many discrepancies were found between LES/SLFM and RANS/SLFM in the prediction of the mean and RMS profiles of temperature and mixture fraction.

B-1 flame:

- Successful predicting of the mean and RMS statistics of temperature, mixture fraction, and species mass fractions.

50

IOWA STATE UNIVERSITY

Concluding Remarks

LES/SLFM

B-1 flame:

- The influence of scalar dissipation rate was clearly observed.
- LES/SLFM produced lower **Scalar dissipation Rate** since it was computed from the resolved scales.
- B-1 flame structure was sensitive to the computation of the scalar dissipation rate .
- H2O mass fraction peak values was not captured accurately in multiple axial locations except for the location x/d=20.

51

IOWA STATE UNIVERSITY

Concluding Remarks

LES/SLFM

B-1 flame:

- The physical location of H2O mass fraction peak values were predicted in all locations.
- The effect of CO2 altered some of the reaction pathways that led to the under-prediction of H2O near the nozzle.
- Downstream the flame, the prediction of all species mean radial profiles improved .

52

IOWA STATE UNIVERSITY

Equations RANS

$$f = \bar{f} + f''$$

$$\tilde{f} = \frac{\overline{\rho f}}{\bar{\rho}}$$

$$\bar{\rho} v''v'' = -2\mu_t \tilde{\bar{\gamma}} - \frac{2}{3}\left(\mu_t \nabla \cdot \tilde{\bar{v}}\right)\bar{\delta} + \frac{2}{3}\bar{\rho}\tilde{k}\bar{\delta}$$

$$\bar{\rho} v''v'' = -2\mu_t \tilde{\bar{\gamma}} - \frac{2}{3}\left(\mu_t \nabla \cdot \tilde{\bar{v}}\right)\bar{\delta} + \frac{2}{3}\bar{\rho}\tilde{k}\bar{\delta}$$

Equations LES

$$\bar{\Phi} = \int_D G(x - x')\Phi(x')dx'$$

$$G(x - x') = \begin{bmatrix} 1, if \left|x - x'\right| \le \dfrac{\Delta}{2} \\ 0, otherwise \end{bmatrix}$$

Dynamic model

$$T_{ij} = \bar{\rho}\tilde{\hat{u}}_i\tilde{\hat{u}}_j - \bar{\rho}u_i u_j \qquad C_s = \frac{L_{ij}}{M_{ij}}$$

$$L_{ij} = \hat{u}_i\hat{u}_j - \tilde{\hat{u}}_i\tilde{\hat{u}}_j$$

$$\tau_{ij}^{SGS} = T_{ij} - \ell_{ij} \qquad M_{ij} = 2\bar{\rho}\Delta^2\left|S\right|S_{ij} - 2\bar{\rho}\Delta^2\left|\tilde{\hat{S}}\right|\tilde{\hat{S}}_{ij}$$

$$C_s = \frac{\left\langle L_{ij}M_{ij}\right\rangle}{\left\langle M_{ij}M_{ij}\right\rangle}$$

Equations SLFM

$$\chi(z) = \frac{a_s}{4\pi} \frac{3\left(\sqrt{\rho_\infty / \rho + 1}\right)^2}{2\sqrt{\rho_\infty / \rho + 1}} \exp\left(-2[erfc^{-1}(2Z)]^2\right)$$

$$a_s = \frac{\chi_{st}\pi}{\exp\left\{-2\left[erfc^{-1}(2Z_{st})\right]^2\right\}}$$

$$\tilde{\chi}_{st} = \frac{\tilde{\chi} f(Z_{st})}{\int_0^1 f(Z)p(Z)dz}$$

$$f(Z) = \exp\left(-2[erfc^{-1}(2Z)]^2\right)$$

55

IOWA STATE UNIVERSITY

491823 Effective Homogenization of Stratified Liquids Using SIDE Entry Mixing

Abstract:

In many industrial applications, effective blending of fluids with different properties is very important. The ability to effectively blend fluids at lower power and torque inputs offers initial (capital) and longer term (power and maintenance) savings. Examining density stratified fluids in petroleum storage tanks is a unique subset of blending. The published literature on this topic is limited, and while newer contributions are available, the bulk of that data is forty to sixty years old.

Blending in oil storage tanks has been accomplished in several different ways historically. This includes the use of side entry mixers, angled or horizontal jet nozzles, and rotating jet nozzles. Although there is evidence that top entry mixers are more effective at blending in this application (Kehn 2011), because of the very large size and geometry of these vessels, conventional top entry mixers are rarely used.

There is strong evidence in the literature that angling a jet such that it is directed from near the base toward the liquid surface can significantly decrease blending and destratification times compared to a horizontal jet (Grenville, Tilton 2011). The use of side entry mixers in angled up orientation has been done for many years in tall tank applications like edible oil storage or in the food and beverage industry. Using a mixer angled up in oil storage has not been done in practice partly because the agitator has two separate duties in the system. In addition to blending, the mixer needs to be able to keep solids and water (basic sedimentation and water, BS&W) off the bottom of the tank.

To optimally meet the requirements of destratification in the upper portions of a vessel and off-bottom suspension at the vessel base the use of both technologies can be useful. Experimental and computational studies comparing blending in the angled and horizontal configuration were performed. This presentation will describe the findings of this study and highlight a new experimental method to examine blending in the presence of temperature based stratification.

Kehn, R. O., "Comparing top entry versus side entry agitator performance in low viscosity blending ", *Can. J. Chem. Eng.*, 89: 1059–1067 (2011)

Grenville, R. K., J. N. Tilton, "Jet mixing in tall tanks: Comparison of methods for predicting blend times", Chemical *Engineering Research and Design*, Volume 89, Issue 12, 2501 – 2506 (2011)

500986 The Influence of Frame Rotation on Heat Transfer in Fully-Developed Turbulent Channel Flows

Abstract:

Turbulent mixing of internal energy near a solid/fluid interface predicted by the URAPS closure for the normalized Reynolds stress is qualitatively consistent with DNS results for non-isothermal rotating channel flows. At high rotation numbers, the dispersion coefficient in the transverse (cross flow) direction is larger than the dispersion coefficient in the spanwise direction. The longitudinal dispersion coefficient is relatively small compared with the transverse and spanwise dispersion coefficients.

495723 Counter-Rotating Vortex Shedding Generated By Acoustic Excitations in Confined Mixing Layers

Abstract:

Novel ultrafast mixing enhancement has been realized in confined mixing layers at an optimal forcing frequency, under which the flow exhibits strongest velocity fluctuations (Wang GR. Chem Eng Sci. 2003, 58:4953-63; Wang, GR, AIChE J, 2006, 52:111-24). Previous investigations showed that, the optimal forcing frequency is attributed to an acoustic resonance which is dominated by the overall geometric scale of water tunnel (Zhao and Wang, Chemical Engineering and Processing: Process Intensification, 111 (2017) 67â€“78). We also found, the fast mixing was always accompanied with the so-called counter-rotating vortices (CRVs). In this investigation, with particle imaging velocimetry (PIV), we quantitatively study the velocity fluctuations in the nozzle section. It is found, under the optimal frequency, the flow exhibits highest velocity fluctuations due to acoustic resonance. We also find, CRVs can be generated are while two conditions are satisfied simultaneously, which are: (1) The acoustic particle displacement must be larger than the summation of the curvature radius of the trailing edge and the thickness of Stokes boundary layer, and (2) the convective transport of mean vorticity of unforced mixing layer should be conquered by the variation of shedding vorticity. These conditions are experimentally validated under different forcing frequencies, Reynolds number of basic flow and velocity ratios etc. Our investigations could be very useful on understanding forced vortex shedding mechanism. The results are also inspiring on designing novel and fast mixer, heat exchanger and sound attenuator etc.

498466 Practical Large Eddy Simulation: Extension toÂ Variable Mesh Systems

Abstract:

A lattice-Boltzmann approach for performing large eddy simulations (LES) across domains with spatially varying lattice resolutions is presented. This approach uses nested sub-volumes to create high-resolution sub-domains about momentum sources inside closed tanks. We begin by discussing the macroscopic and microscopic communication required to link lattices with disparate resolutions. Next, we discuss the computational overhead associated with nested sub-volumes, as opposed to conventional approaches using a uniform lattice. For systems with large tank-to-impeller diameter ratio, the variable mesh approach is shown to be order-of-magnitude faster. We conclude with a set of validation exercises, specific to mixing and agitator performance.

503577 Species Exposure As an Approach to Quantifying Mixer Performance

Abstract:

While bulk uniformity is easily determined using blendtime, experimentally it lacks the ability to describe local blending behavior. As an alternative to uniformity, the so-called "exposure" of secondary materials in a bulk material provides both local and bulk mixing characteristics. Exposure, as defined by Kresta et al., relates the contact area between multiple species, which allows mass and heat transfer to be analyzed through a similar metric [1]. It also unifies the analysis of rotating systems and static mixer systems with a universal mixing description. While the mathematical viability of this parameter has been studied, it has yet to be fully analyzed with the context of a time-varying, three-dimensional mixing system. Validation of exposure in both rotating and static mixed systems is presented here and the viability of it as a universal mixing descriptor is discussed.

[1] S. M. Kresta, A. Kukukova, J. Aubin. "A new definition of mixing and segregation: Three dimensions of a key process variable". Chemical Engineering Research and Design 87 (2009) pp. 633-647

499648 Large Eddy Simulation of a Pipeline Rotor-Stator Mixer

Abstract:

Rotor-stator mixers provide high deformation rates to a relatively limited volume, resulting in a region in which intensive mixing, milling, and/or dispersion/emulsification operations can occur. Because they are geometrically complex, experimental measurement of important flow regions within rotor-stator mixers is often difficult, if not impossible. Computational fluid dynamics (CFD) simulations of mixers are a means to provide flow field information that are of benefit to both designers and end users.

This presentation focuses on our recent work in the fully transient (sliding mesh) three-dimensional large eddy simulation (LES) of a pipeline rotor-stator mixer, the Greerco High Shear Pipeline Mixer, using ANSYS FLUENT software. The modeled unit consists of two conical rotor-stator stages, each with a nominal diameter of 4 inches, aligned to produce an axial discharge flow. The mixer geometry consists largely of cylindrical and conical bodies; the impact of this on meshing will be discussed. The working fluid was water in turbulent flow. The wall-adapting local eddy-viscosity model is used to model subgrid scales and an equilibrium stress wall model describes the interaction of scales near no-slip surfaces.

Flow and turbulence quantities have been studied on a per stator slot and per rotor stage basis respectively. As with other rotor-stator mixers, fluid is shown to impinge on the trailing surface of the stator slot. Turbulent kinetic energy and its isotropic dissipation rate are shown in such regions. Comparisons are made between the LES filtered Navier-Stokes simulations and our previously reported Reynolds averaged Navier-Stokes (RANS) realizable k-Îµ model simulations. Turbulence scales important for mixing are approximated and compared for both of these turbulence models. The effect of mesh density to resolve the larger scales of turbulence are discussed by comparing results from a coarse and refined computational mesh. The effect of operating conditions on power draw, throughput and other quantities of practical utility are also discussed.

501402 Mean Age Distribution in General Time Dependent Flows of Multiphases

Abstract:

Residence time and age analysis has been a standard tool in chemical engineering for many types of flow systems. In this talk, governing equations for mean age and higher moments for general fluid flows of time dependent, multispecies, multiphase, with arbitrary number of inlets and outlets will be discussed. Tracer species can be released at any inlets or selected locations in the interior of the flow. It will be shown that mean residence time of a species or phase at an outlet is the mass flow averaged mean age at the outlet. Residence time frequency function can be defined at an outlet with the age frequency function that defines mean age in the interior of a flow. This residence time frequency function extends current residence time theory to fully time dependent for multiple species and multiple inlets and outlets. For a steady flow, the sum of the mass flow weighted mean residence times of a species at all outlets is equal to the total mass of the species in the flow. For an unsteady flow, mean residence time also depends on the rate of change of mass averaged age concentration. With these equations mean age and residence time analysis can be extended to flows and mixing in all types of flow systems, not limiting to steady flows of single phase only anymore.

496416 Calibration of Mean Age for Blend Time in Batch Processes

Abstract:

Applying mean age theory to calculate blend times using CFD simulations in stirred vessels promises to significantly reduce overall computation time by eliminating the need of steady state simulations. For continuous vessels, the application of this theory is straight forward. For batch vessels, an inlet and outlet with an arbitrarily low flow rate are added to the simulation in order to satisfy mean age boundary conditions. This method however needs to be verified for various operating configurations of stirred vessels. In this follow-up to our 2016 study, calibration of inlet and outlet locations is examined. Impeller diameter, impeller clearance and overall scale are varied to determine the consistency of the calibrated locations. Mean age results will be compared against traditional transient results and available experimental data. General rules for calibrating mean age simulations in batch process will be presented. All simulations were completed using STAR-CCM+.

485435 Modeling the Residence Time Distribution inÂ a Continuous Plug-Flow Reactor

Abstract:

In a plug-flow reactor, the residence time distribution (RTD) directly impacts catalyst use, reaction selectivity and reaction yield. The RTD can be measured by tracer studies, but such studies are expensive. We investigate the residence time distribution in a 200 cubic meter industrial-scale plug-flow reactor composed of 13 agitated compartments using both RANS CFD simulations and LES Lattice Boltzmann method (LBM) simulations, and compare the results with tracer measurements. We calculate residence time distributions with mean-age theory in RANS models and with both mean-age theory and particle tracking in LBM models. As the particle based methods have no numerical diffusion, this allows us to quantify the effects of numerical diffusion in mean-age theory. Mean-age theory provides not only the residence time distribution, but also insight into dead-zones; it is not, however, readily measured. Residence time distribution is readily measurable by tracer studies but provides no local information on mixing. Simulation models provide an inexpensive mans of understanding local mixing and overall residence time distributions, as well as providing methods of screening process modifications to enhance the residence time distribution.

494947 AnalyzingÂ Direct Numerical SimulationÂ Data forÂ a Marginally Turbulent Stirred Vessel Driven By a Rushton Turbine

Abstract:

New results will be presented from a Direct Numerical Simulation (DNS) of the just turbulent flow field, at Re=7300, in a stirred vessel provided with baffles and driven by a Rushton turbine. The simulations have been carried out by means of a Lattice Boltzmann solver, written in C/CPP, on a lattice comprising some 30 billion nodes, carried out at the Irish Centre for High-End Computing ICHEC in Dublin.

The results of this simulation are compared with experimental data from the literature and with the results of the less refined DNS reported earlier. The main variables of interest are the turbulent kinetic energy and its dissipation rate, phase averaged as well as phase resolved, which can be calculated from the outcome of such a DNS at any position in the vessel. Also turbulent spectra can be obtained in this way and some spectra will be presented for some positions in the vessel.

The idea behind performing DNS for stirred vessel is to provide a yardstick for assessing the findings of Large Eddy Simulations and for evaluating the performance of sub-grid scale models essential for such LESs. Even experimental data may be assessed by means of a DNS as the resolution attainable in DNSs is better than in experiments.

500011 A New Mixing Model for Turbulent Reacting Flows Using Hierarchical Parcel Swapping (HiPS)

Abstract:

Turbulent reacting flows are ubiquitous in Chemical Engineering and a great deal of work has been performed to understand and quantify them. This is an extremely challenging problem, however, due to the complexity of turbulent flows. Such flows exhibit a wide range of length and time scales. Resolving all continuum scales is not possible for engineering applications due to large Reynolds numbers. The computational cost of direct simulations that do resolve all scales increases with the cube of the Reynolds number. Large Eddy Simulations (LES) overcome this restriction by only resolving the large scale eddies, while modeling the subgrid scale mixing and reaction processes. The key challenge for LES is developing accurate subgrid mixing models, and there are a number of models of varying success that have been published. These include the Interaction by Exchange with the Mean (EIM) model, the Euclidian Minimum Spanning Trees (EMST) model, Curl's model, and the Shadow Position Mixing Model. The central challenge of these models is representing the mixing processes in a manner that is computationally efficient while retaining as much physics as possible. We present a new mixing model termed Hierarchical Parcel Swapping (HiPS). HiPS was recently introduced and applied to channel flow in two publications. HiPS consists of a collection of notional fluid parcels with adjacency defined using a binary tree. The levels of the tree correspond to geometrically-spaced length scales, the number of which are user-defined. The mixing and pairing of parcels is done at rates consistent with inertial range turbulent scaling, and this is the key physical processes enabling the potential improvement of HiPS over other models. The model has aspects in common with the Linear Eddy and One Dimensional Turbulence models, but is more efficient and flexible, and does not directly imply diffusive mixing on a physical domain. We present an overview of the HiPS model, and then extend the model to reacting flows. Reaction adds complexity to the mixing process through variable properties (including dilatation associated with heat release), the need to carry multiple scalars, which may have varying Schmidt numbers, and the development of complex reaction-transport structures, such as flames. At this early stage of HiPS development, we demonstrate the model for simple parallel competitive reaction mechanisms. We consider variable Schmidt number and dilatational effects. Results are presented at varying Damkohler and Reynolds numbers. Product selectively is discussed.

490866 Application of Computational Fluid Dynamics to Assess Scale-up Risk to Manufacture Polymeric Resins for the Paints and Coatings Industry

Abstract:

The Valspar Corporation manufactures polymeric resins for use in the paints and coatings industry. Evaluating the mixing and heat transfer requirements is important to de-risk the scale-up of new resin synthesis processes or transferring synthesis processes between existing manufacturing assets. Several computational fluid dynamic (CFD) tools have been applied to evaluate critical steps associated with mixing and heat transfer. In this talk, we will discuss the application of CFD tools to evaluate the risk of scaling up a new resin synthesis process and to transfer an existing synthesis process between two assets.

MODELING OF DROPLET COALESCENCE AND BREAKAGE IN OIL-WATER PIPELINES USING ONE-GROUP INTERFACIAL AREA TRANSPORT

Eulerian model used to model oil water flow regimes using Interfacial Area Concentration model in ANSYS Fluent

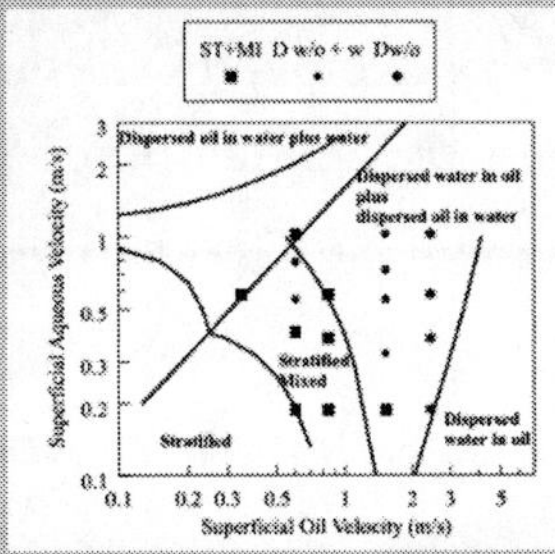

Kuochen Tsai
Shell International Exploration & Production Inc.

DEFINITIONS & CAUTIONARY NOTE

Reserves: Our use of the term "reserves" in this presentation means SEC proved oil and gas reserves.

Resources: Our use of the term "resources" in this presentation includes quantities of oil and gas not yet classified as SEC proved oil and gas reserves. Resources are consistent with the Society of Petroleum Engineers 2P and 2C definitions.

Organic: Our use of the term Organic includes SEC proved oil and gas reserves excluding changes resulting from acquisitions, divestments and year-average pricing impact.

Shales: Our use of the term 'shales' refers to tight, shale and coal bed methane oil and gas acreage.

The companies in which Royal Dutch Shell plc directly and indirectly owns investments are separate entities. In this document "Shell", "Shell group" and "Royal Dutch Shell" are sometimes used for convenience where references are made to Royal Dutch Shell plc and its subsidiaries in general. Likewise, the words "we", "us" and "our" are also used to refer to subsidiaries in general or to those who work for them. These expressions are also used where no useful purpose is served by identifying the particular company or companies. "Subsidiaries", "Shell subsidiaries" and "Shell companies" as used in this document refer to companies over which Royal Dutch Shell plc either directly or indirectly has control. Companies over which Shell has joint control are generally referred to as "joint ventures" and companies over which Shell has significant influence but neither control nor joint control are referred to as "associates". The term "Shell interest" is used for convenience to indicate the direct and/or indirect ownership interest held by Shell in a venture, partnership or company, after exclusion of all third-party interest.

This presentation contains forward-looking statements concerning the financial condition, results of operations and businesses of Royal Dutch Shell. All statements other than statements of historical fact are, or may be deemed to be, forward-looking statements. Forward-looking statements are statements of future expectations that are based on management's current expectations and assumptions and involve known and unknown risks and uncertainties that could cause actual results, performance or events to differ materially from those expressed or implied in these statements. Forward-looking statements include, among other things, statements concerning the potential exposure of Royal Dutch Shell to market risks and statements expressing management's expectations, beliefs, estimates, forecasts, projections and assumptions. These forward-looking statements are identified by their use of terms and phrases such as "anticipate", "believe", "could", "estimate", "expect", "intend", "may", "plan", "objectives", "outlook", "probably", "project", "will", "seek", "target", "risks", "goals", "should" and similar terms and phrases. There are a number of factors that could affect the future operations of Royal Dutch Shell and could cause those results to differ materially from those expressed in the forward-looking statements included in this presentation, including (without limitation): (a) price fluctuations in crude oil and natural gas; (b) changes in demand for Shell's products; (c) currency fluctuations; (d) drilling and production results; (e) reserves estimates; (f) loss of market share and industry competition; (g) environmental and physical risks; (h) risks associated with the identification of suitable potential acquisition properties and targets, and successful negotiation and completion of such transactions; (i) the risk of doing business in developing countries and countries subject to international sanctions; (j) legislative, fiscal and regulatory developments including potential litigation and regulatory measures as a result of climate changes; (k) economic and financial market conditions in various countries and regions; (l) political risks, including the risks of expropriation and renegotiation of the terms of contracts with governmental entities, delays or advancements in the approval of projects and delays in the reimbursement for shared costs; and (m) changes in trading conditions. All forward-looking statements contained in this presentation are expressly qualified in their entirety by the cautionary statements contained or referred to in this section. Readers should not place undue reliance on forward-looking statements. Additional factors that may affect future results are contained in Royal Dutch Shell's 20-F for the year ended 31 December, 2015 (available at www.shell.com/investor and www.sec.gov). These factors also should be considered by the reader. Each forward-looking statement speaks only as of the date of this presentation, 14 November 2016. Neither Royal Dutch Shell nor any of its subsidiaries undertake any obligation to publicly update or revise any forward-looking statement as a result of new information, future events or other information. In light of these risks, results could differ materially from those stated, implied or inferred from the forward-looking statements contained in this presentation. There can be no assurance that dividend payments will match or exceed those set out in this presentation in the future, or that they will be made at all.

We use certain terms in this presentation, such as discovery potential, that the United States Securities and Exchange Commission (SEC) guidelines strictly prohibit us from including in filings with the SEC. U.S. Investors are urged to consider closely the disclosure in our Form 20-F, File No 1-32575, available on the SEC website www.sec.gov. You can also obtain this form from the SEC by calling 1-800-SEC-0330.

INTRODUCTION

- Understanding water-oil separation in pipelines is critical for aqueous corrosion management.
- Corrosion can cause serious production loss and environmental contamination.
- Similar models have been well studied for gas/liquid flows and the same mechanism for breakage and coalescence can be borrowed.
- One-group population model has a good balance between model complexity and computational costs.
- Recent advancement in oil/water flow measurement allows more accurate model validation.

CFD MODEL – DROPLET TRANSPORT

- CFD model
  - Eulerian with interfacial area concentration (IAC)
  - IAC defined as the ratio between water droplet area/water droplet volume
  - Relation: $IAC = \frac{6\times Vol\ Frac_{water}}{Droplet\ Diameter}$ assuming spherical droplets
  - Breakage and coalescence model: Ishii and Kim (2001)*
  - Transport equation of IAC (χ):

$$\frac{\partial\left(\rho_g\chi_p\right)}{\partial t}+\nabla\cdot\left(\rho_g\overline{u}_g\chi_p\right)=\frac{1}{3}\frac{D\rho_g}{Dt}\chi_p+\frac{2}{3}\frac{\dot{m}_g}{\alpha_g}\chi_p+\rho_g(S_{RC}+S_{WE}+S_{TI})$$

  - Where ρ is density, α volume fraction, S_{RC}, S_{WE} and S_{TI} are the source term for coalescence by collision, wake entrainment and breakage due to turbulence shearing

- *Two phase model by Ishii and Kim (*Nuclear Engineering and Design* **205**, 2001)

CFD MODEL – DROPLET COALESCENCE AND BREAKAGE

- Coalescence and breakage functional form

$$S_{RC} = -\frac{1}{3\phi}\left(\frac{\alpha_g}{X_p}\right)^2 C_{RC}\left[\frac{n_p^2 u_t d_b^2}{\alpha_{gmax}^{1/3}\left(\alpha_{gmax}^{1/3}-\alpha_g^{1/3}\right)}\right]\left[1-exp\left(-C\frac{\alpha_{gmax}^{1/3}\alpha_g^{1/3}}{\alpha_{gmax}^{1/3}-\alpha_g^{1/3}}\right)\right]$$

$$= -\frac{1}{3\pi} C_{RC} u_t X_p^2\left[\frac{1}{\alpha_{gmax}^{1/3}\left(\alpha_{gmax}^{1/3}-\alpha_g^{1/3}\right)}\right]\left[1-exp\left(-C\frac{\alpha_{gmax}^{1/3}\alpha_g^{1/3}}{\alpha_{gmax}^{1/3}-\alpha_g^{1/3}}\right)\right]$$

$$S_{WE}=-\frac{1}{3\phi}\left(\frac{\alpha_g}{X_p}\right)^2 n_b^2 d_b^2 u_r C_D^{1/3}=-\frac{1}{3\pi}C_{WE}u_r X_p^2 C_D^{1/3}$$

$$S_{TI} =\frac{1}{3\phi}\left(\frac{\alpha_g}{X_p}\right)^2 C_{TI}\left(\frac{n_b u_t}{d_b}\right)\left(1-\frac{We_{cr}}{We}\right)^{1/2} exp\left(-\frac{We_{cr}}{We}\right)$$

$$=\frac{1}{18}C_{TI}u_t\frac{X_p^2}{\alpha_g}\left(1-\frac{We_{cr}}{We}\right)^{1/2} exp\left(-\frac{We_{cr}}{We}\right)$$

$u_t = \varepsilon^{1/3} d^{1/3}$ Bubble fluctuating vel

$u_r=\left(\frac{d_b g\Delta\rho}{3C_D\rho_f}\right)^{1/2}$ Bubble terminal vel

$C_D=24\frac{\left(1+0.1Re_D^{0.75}\right)}{Re_D}$ and $Re_D\equiv\frac{\rho_f u_r d_b}{\mu_f}(1-\alpha_g)$

$We=\frac{\rho_f u_t^2 d_b}{\sigma}$

- Effect of viscosity is incorporated in drag law through Reynolds number
- Effects of surface tension is included through the Weber number
- Either viscosity or surface tension impacts coalescence in this model, but still can be included by modifying C_{RC} since both are fluid properties independent of flow variables
- The above correlations are geometrical and account for collision frequency only. Collision efficiency effects are discussed next

*Two phase model by Ishii and Kim (*Nuclear Engineering and Design* 205, 2001)

CFD MODEL – DROPLET COALESCENCE EFFICIENCY

- Coalescence efficiency formula*:

$\lambda(d_1,d_2) = exp\{-\phi(We/2)^{1/2}\}$ - *inertial* if $d_{12}/\eta \gg 1$

$\lambda(d_1,d_2) = exp\{-\phi d_{12}\mu_d Ca^{1.5}/(\mu h_f)\}$ - viscous if $d_{12}/\eta \ll 1$

Where

$1/d_{12} = 0.5(1/d_1 + 1/d_2)$

$Ca = \mu(\varepsilon/v)^{0.5} d_{12}/2\sigma$ – *Capillary number*

$h_f = (0.6\times10^{-20} d_{12}/16\pi\sigma)^{1/3}$ - *Critical film thickness*

$\eta = (v^3/\varepsilon)^{1/4}$ – *Kolmogorov length scale*

- Correlations based the work of Chesters (1991)
- Inertial formulation for fully turbulent flows, viscous one for laminar flows
- Inertial formulation shows weak dependency on fluid properties, thus can be assumed constant for our applications.
- In contrast, the laminar formula shows high sensitivity to fluid viscosity and surface tension and can vary by several orders of magnitude with small changes in fluid properties, an unlikely scenario for oil industry
- Consistent with other studies shown in the review work of Liao and Lucas (*CES 65*, 2010)

*Chesters, A K, The modeling of coalescence processes in fluid-liquid dispersion: a review of current understanding. Trans. *IChemE 69 (Pt. A)*. 1991

EXPERIMENTAL DATA

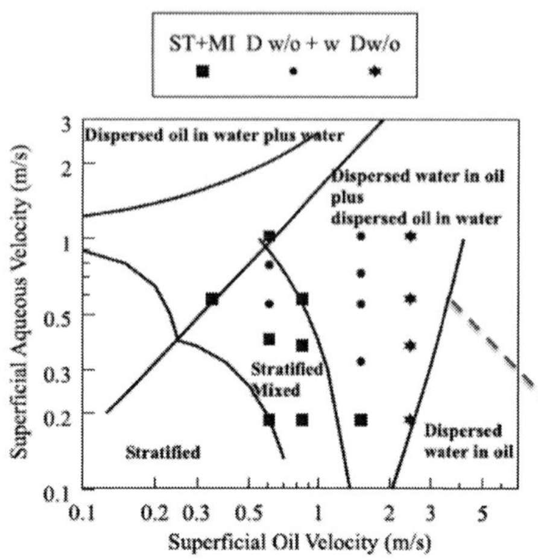

- Fluid properties:
  - Aqueous: density -1.116 SG, viscosity – 1.6 cp
  - Oil: density – 0.797 SG, viscosity – 1.8 cp
  - Interfacial tension: 0.01 N/m
- Flow configurations:
  - Pipe ID: 0.064 m
  - Pipe length: 4.5 m
  - Dispersed phase vol. %: 6-42

Trallero map over-predicts dispersed velocity location at high aqueous content

- Experimental data from Simmons and Azzopardi (IJFM 27, 2001)*
- Flow map by Trallero (PhD Thesis, Tulsa U., 1995)**

MODEL VALIDATIONS

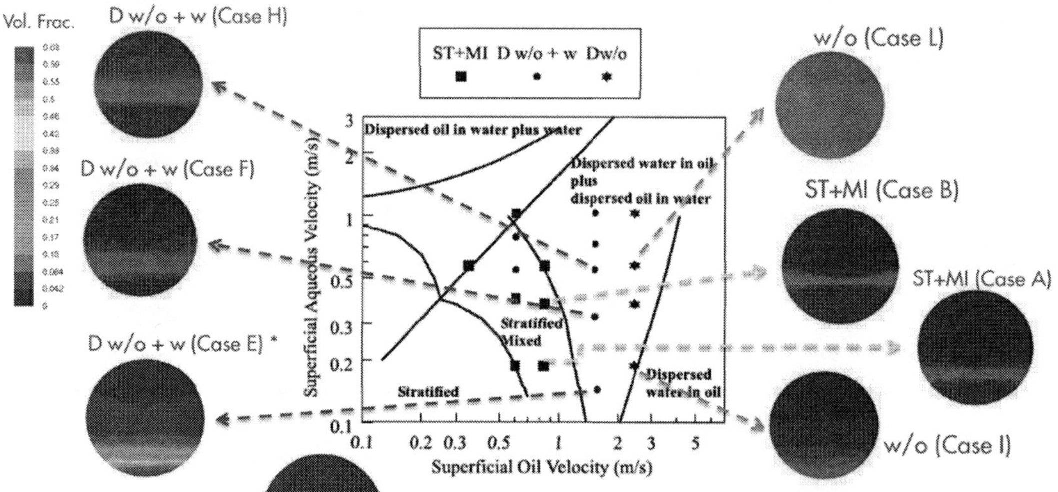

ST – stratified
MI - Mixing
D w/o + w – dispersed water in oi plus water layer
* Run with Quick Scheme for both VF and IAC

- Dispersed, mixing and stratified water are well reproduced!
- CFD model shows better prediction accuracy than Trallero flow map
- C_{RC} = 0.004, C_{WE}=0.002, C_{TI}=16000 (Default 0.085), We_{cr} = 0.23 (6.0 for air/water)

CFD MODEL FLOW CONDITIONS SUMMARY

Kerosene superficial Velocity U_{so} (m/s)	Aqueous phase superficial velocity U_{sw}	Mixture superficial velocity U_{mix}	Dispersed phase concentration (% vol)	Horizontal flow pattern	Case Number	Flow pattern determined from model
0.837	0.158	0.995	15.9	ST+MI	A	ST+MI
0.837	0.317	1.154	27.5	ST+MI	B	ST+MI
0.837	0.488	1.325	36.8	D w/o+w	C	
0.837	0.614	1.451	42.3	D w/o+w	D	
1.49	0.158	1.648	9.6	D w/o+w	E	D w/o+w
1.49	0.317	1.807	17.5	D w/o+w	F	D w/o+w
1.49	0.488	1.978	24.7	D w/o+w	G	
1.49	0.614	2.104	29.2	D w/o+w	H	D w/o+w
2.393	0.158	2.551	6.2	w/o	I	w/o
2.393	0.317	2.71	11.7	w/o	J	
2.393	0.488	2.881	16.9	D w/o+w	K	
2.393	0.614	3.007	20.4	w/o	L	w/o

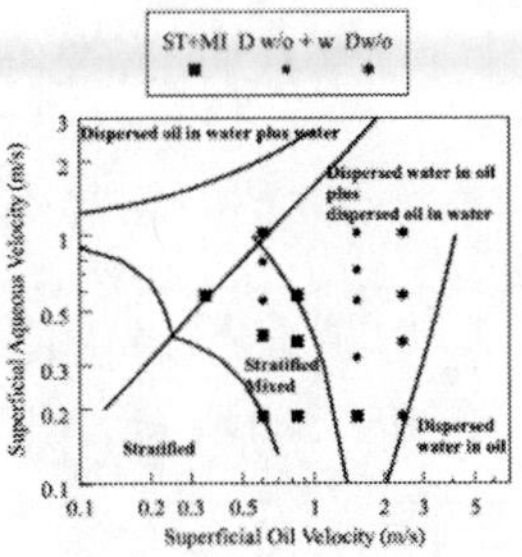

Stratified Flow				
U_{so}	U_{sw}	h/D	Case	Predicted h/D
0.84	0.158	0.28	A	0.23
0.84	0.317	0.32	B	0.34
0.84	0.488	0.34	C	
0.84	0.614	0.53	D	
1.49	0.158	0.24	E	0.12
1.49	0.317	0.28	F	0.22
1.49	0.488	0.33	G	
1.49	0.614	0.39	H	0.31
1.49	0.88	0.42	M	

- Similar water layer height observed from model results in previous slide

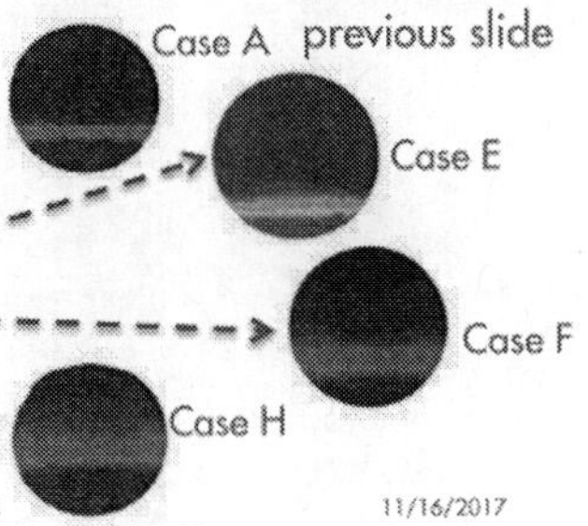

DROPLET SIZES

Droplet size sampling locations

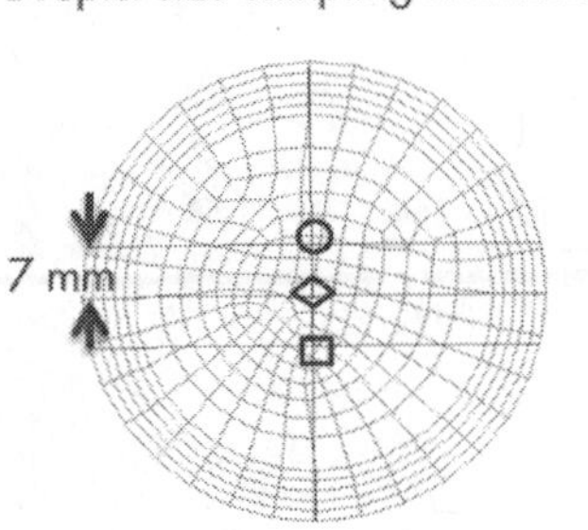

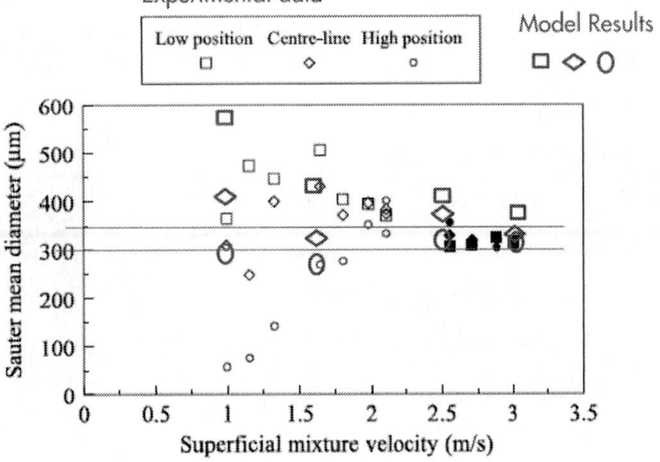

- Particle size distribution approximately matches the experimental data
- The scatter is significant and the results only shows diameter data averaged over the last 0.5 meter section at the specified heights

COMPARISONS WITH FULL-SCALE FAIRUZOV EXPERIMENTAL RESULTS

16" pipe 11200" long pipe (700 L/D)

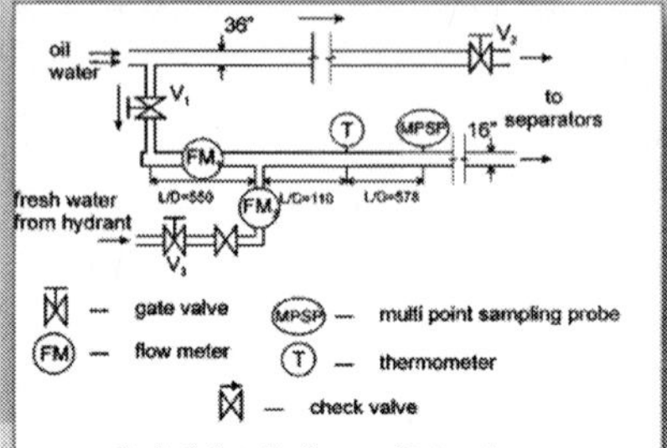

Fig. 3 Schematic diagram of test section

FLUID PROPERTIES AND EXPERIMENTAL CONFIGURATIONS

- Oil
  - Density: 0.851 SG
  - Viscosity: 4.84 cp @100 °F
- Water
  - Density: 1 SG
  - Viscosity: 0.65 cp @ 100 °F
  - Interfacial surface tension: 0.03 n/m
- Pipe ID: 16"
- Pipe length: 284 m

Experimental Configurations – branching out from a 36" production crude oil production line

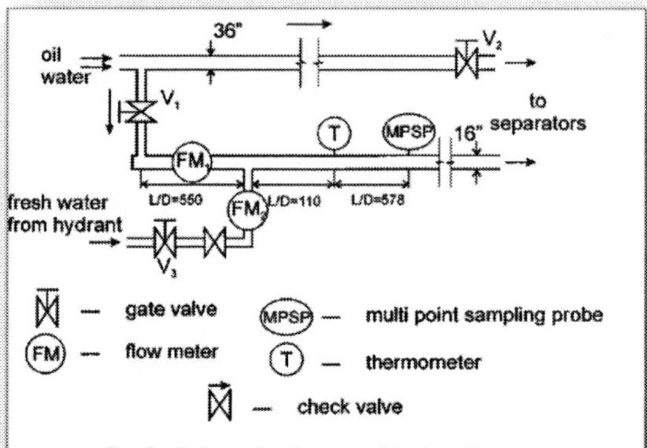

Fig. 3 Schematic diagram of test section

- Y. Fairuzov, P. Arenas-Medina, J. Verdejo-Fierro and R. Gonzalez-Islas, "Flow pattern transitions in horizontal pipelines carrying oil-water mixtures: full-scale experiments," Journal of Energy Resources Technology, **122**, 2000, pp. 169-176

EXPERIMENTS VS CFD MODEL – 3% WATER CUT

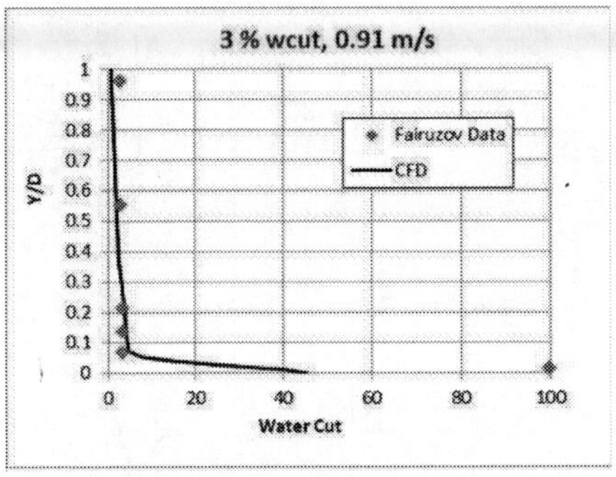

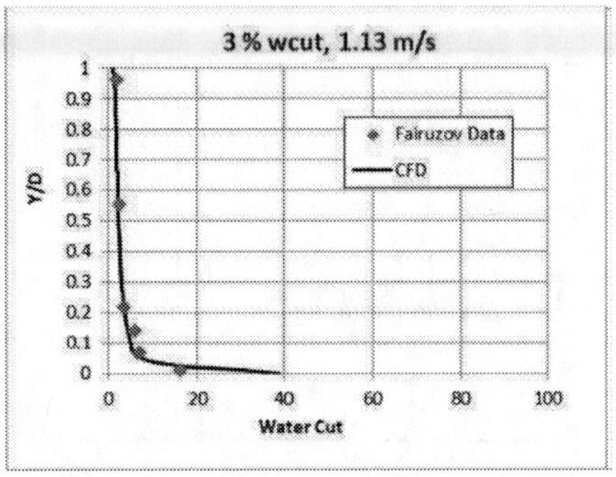

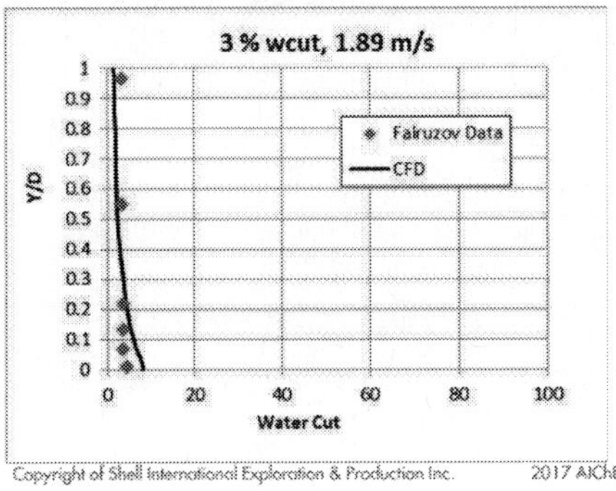

- Excellent match except at bottom when water can form a 100% stratified layer
- However at 3% the thin water layer at pipe bottom is beyond the model resolution

EXPERIMENTS VS CFD MODEL – 7~ 8% WATER CUT

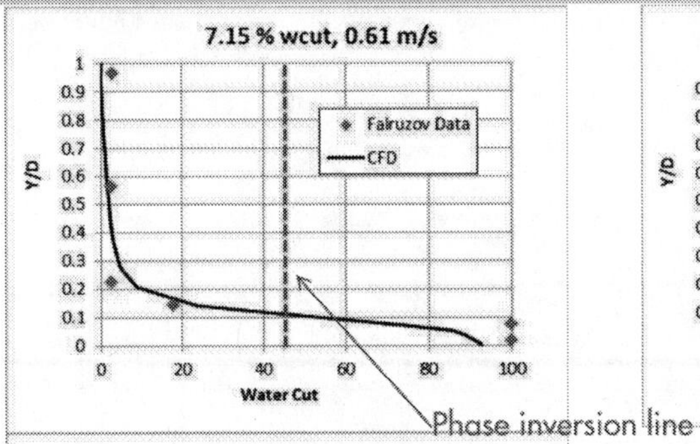

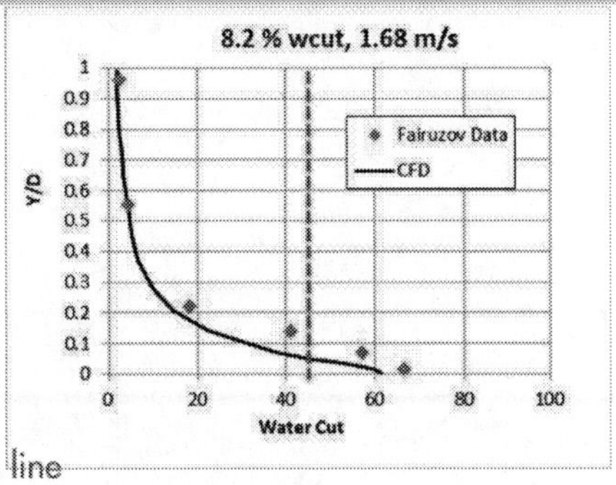

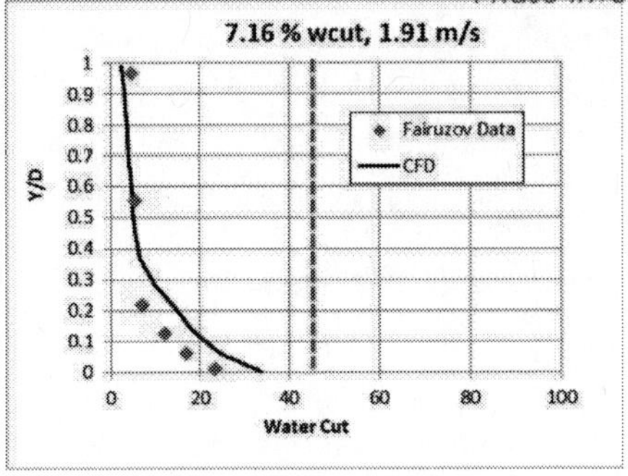

- Excellent match except at bottom when water can form a 100% stratified layer
- 90% of water cut in model shows good match for stratified flow

FLOW REGIME MAP

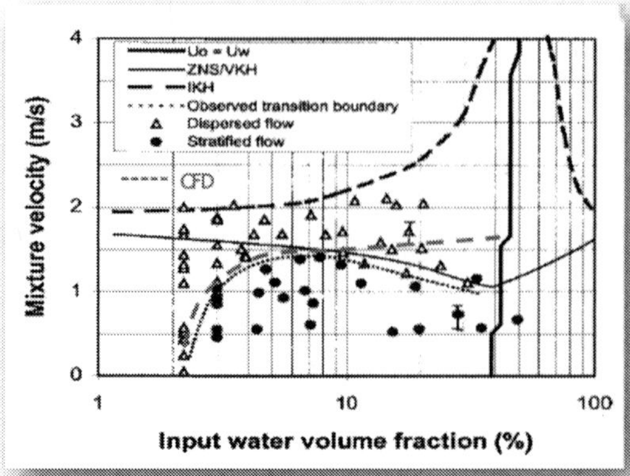

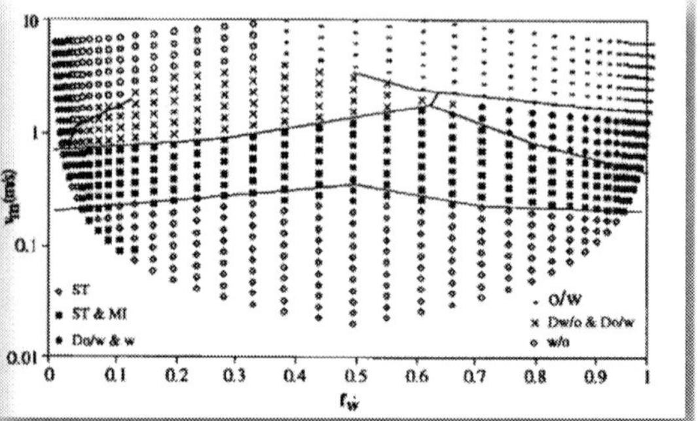

- CFD model predicts well at water cuts < 10%
- At high water cut, CFD model tends to over-predict the stratified flow regime, indicating the functional form of droplet coalescence may need modification with lower dependency on water volume fraction or extra physics, e.g. emulsion, need to be included.
- However, such behavior was not observed in Trallero's flow map shown on the right. The ST&MI region show a positive slope with water cut up to the 60% level and only declines beyond the phase inversion where water becomes the continuous phase. (viscosity ratio ~30, surface tension 36 mN/m)

COMPARISONS WITH DIRECT WATER CONTACT MEASUREMENT OF OHIO U.

Most recent data from Ohio U. showed direct measurement of water wetting along the bottom of a horizontal pipe using electrodes, allowing a more precise measurement of water wetting for validation

BENCHMARK CASES – DIRECT MEASUREMENT OF WATER WETTING*

4 velocities:

- 0.5 m/s
  - 3, 5 and 20%
- 1.0 m/s
  - 1(not shown in photos), 5, 10 and 18%
- 1.5 m/s
  - 1, 5 and 10%
- 1.7 m/s
  - 1 and 5%

- Pipe ID: 0.1 m
- Oil: 0.823 SG, 2.7 cP, 28.5 mN/m
- Water: 1.0 SG, 1 cP, 72 mN/m
- Surface tension oil/water: 40.5 mN/m
- Water-in-oil contact angle: 73°
- Initial droplet size: 1mm

**K.E. Kee, S. Richter, M. Babic and S. Nesic,
"Experimental study of oil-water Flow patterns in a large diameter flow loop – the effects on water wetting and corrosion," Corrosion, 72 (4), 2016

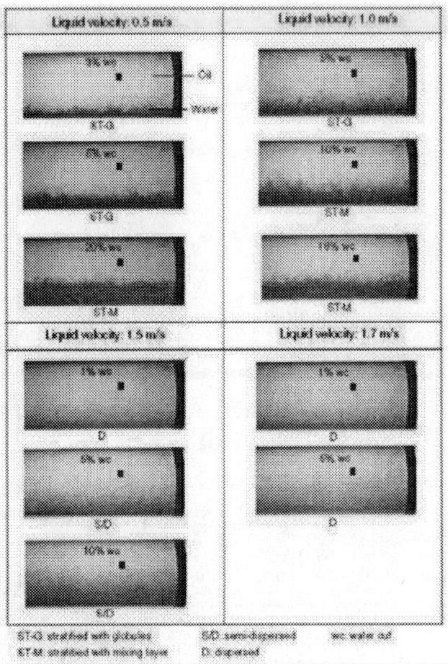

FIGURE 6. Images of horizontal oil-water flow patterns taken by high-speed camera for liquid velocities V_m from 0.5 m/s to 1.7 m/s and water cuts from 5% to 20%.

SURFACE WETTING MAP BY CFD

--- Pipe bottom
--- Pipe center

1 m/s, 10% wcut WW

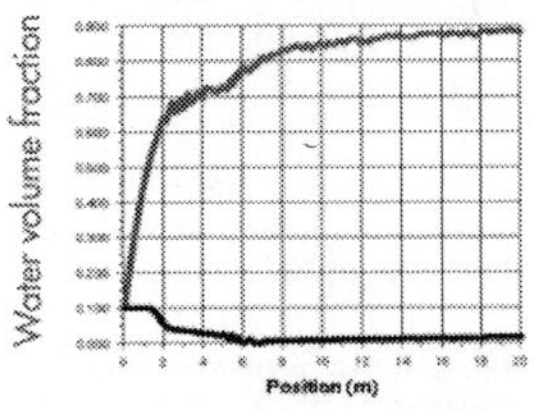

1.5 m/s, 5% wcut –WW/SD

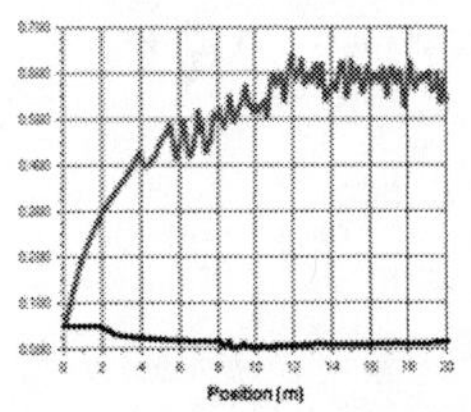

1 m/s, 3% wcut WW

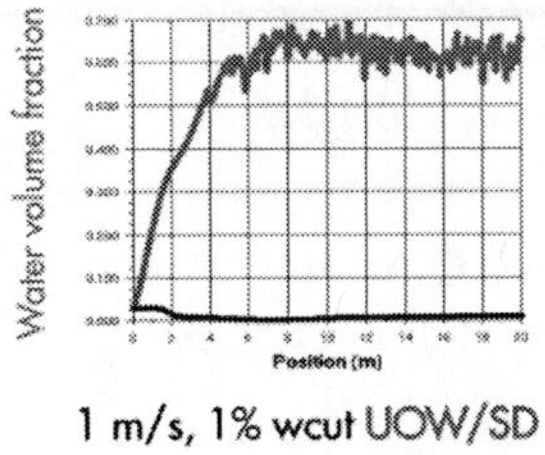

1.5 m/s, 3% wcut –UOW/SD

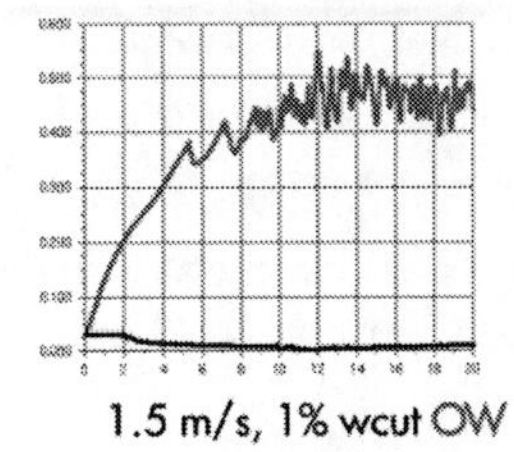

1.7 m/s, 5% wcut UWW

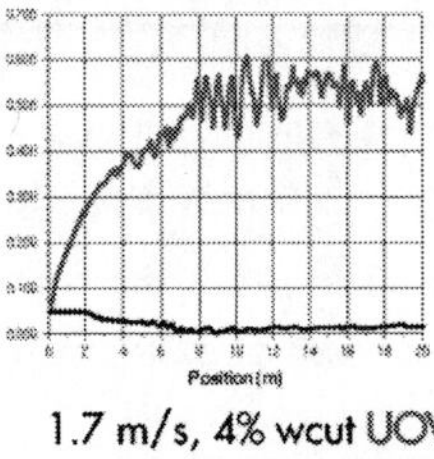

1 m/s, 1% wcut UOW/SD

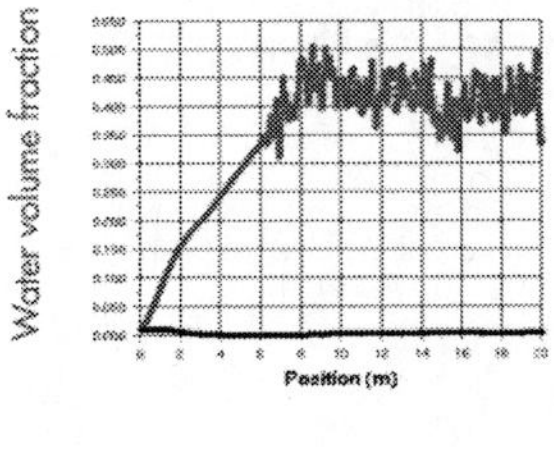

1.5 m/s, 1% wcut OW

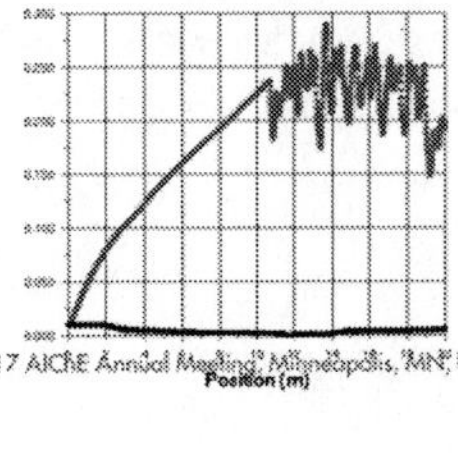

1.7 m/s, 4% wcut UOW

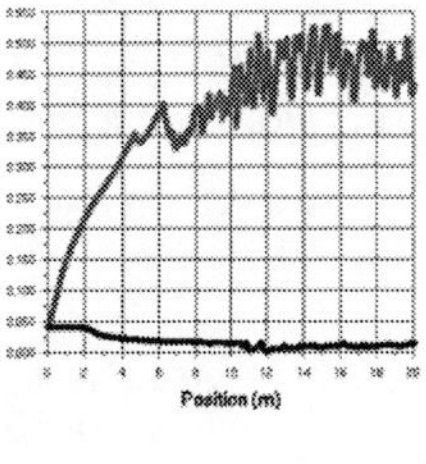

SURFACE WETTING MAP BY OHIO U. – SURF TEN 0.07 N/M

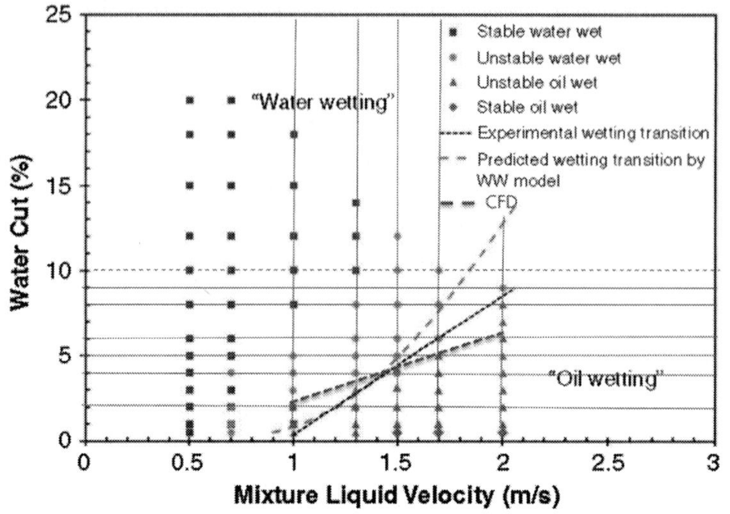

FIGURE 9. *Surface wetting map for horizontal light refined paraffinic oil water flow using new pins design.*

 2017 AIChE Annual Meeting, Minneapolis, MN, USA 11/16/2017 19

CONCLUSIONS

- A one-group interfacial area transport CFD model for oil-water pipeline flows is developed and tuned against lab-scale data, and successfully scaled up to full-scale field data.
- Breakage and collision models are critical:
  - The importance of coalescence is addressed through parametric studies and can be assumed to be constant within the range of parameters of interest.
  - Droplet breakage is critical in validating droplet size distribution and is found to have a value several orders higher than the original model proposed for gas/liquid flows.
- With breakage and coalescence model tuned for droplet size data, scale-up to field data with 16″ ID pipe showed excellent results up to 10% water cuts.
- CFD model uses the full flow history of water droplets in the pipeline to predict water's separation from the oil phase in contrast to experimental correlations where averaged flow history is used.
- CFD can account for more complex geometry such as pipe inclination, bends and turns, etc., thus is more suitable for general purpose use.

 2017 AIChE Annual Meeting, Minneapolis, MN, USA 11/16/2017 20

APPENDIX SLIDES

Slides not used in the main presentation but can be useful for discussions.

SIMULATION NOTES

- Numerical accuracy is important – make sure the mass flux is conserved.
  - Long pipeline requires slower convergence to reach a stationary state, thus if mass flux is off then the results can significantly deviate from the true solution.
  - Convergence can be slow, so even if the residual shows no further change, flux conservation must be ensured.
- Droplet size calculation must be consistent, meaning that if there is no breakage and aggregation then droplet size should remain the same.
  - This is not guaranteed by the numerical algorithm.
  - Consistent method must be chosen between phase and area concentration.
- Wall shear stress value is critical as well – use scalable wall function to avoid error.

IMPACT OF DRAG MODEL

Universal Drag vs. Symmetric Drag (Case E)

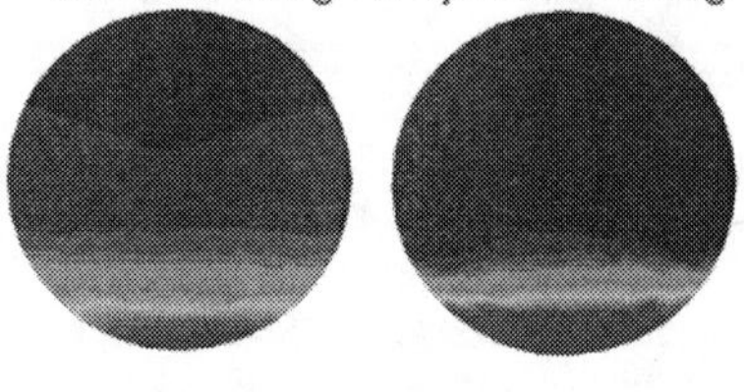

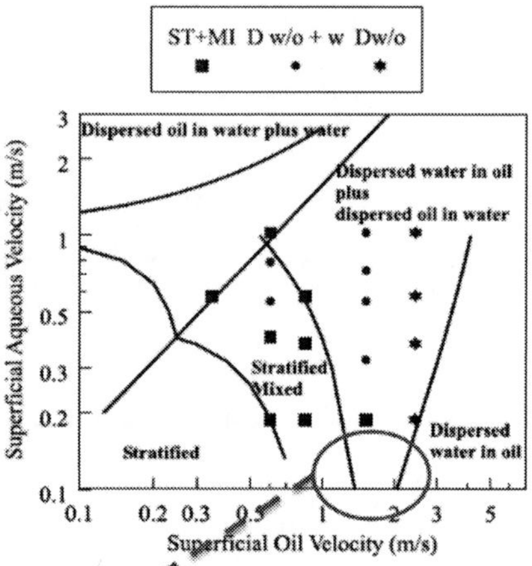

- Drag impact more significantly for stratified flows where continuous phase can switch at the pipe bottom
- Symmetric drag provides a better solution than the Universal drag law because the Symmetric law accounts for phase switch
- However, Symmetric drag still under-estimate the water accumulation at pipe bottom when both the mixture velocity and water cut are low (<10 vol%)
- Both model predicts similar results in dispersed conditions
- Model is less accurate in the transition zone between stratified and dispersed flow regimes, especially where the transition zone is narrow

EXPERIMENTS VS CFD MODEL – 18-~20% WATER CUT

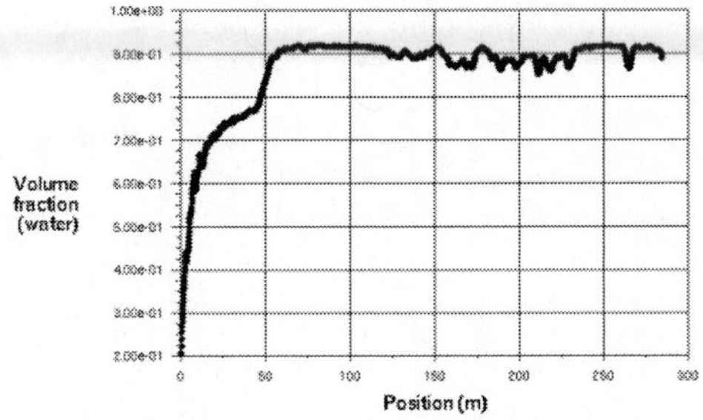

Reaches steady-state at ~60 m

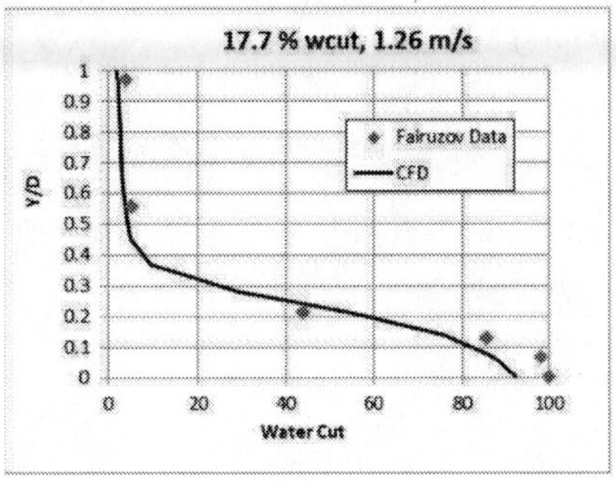

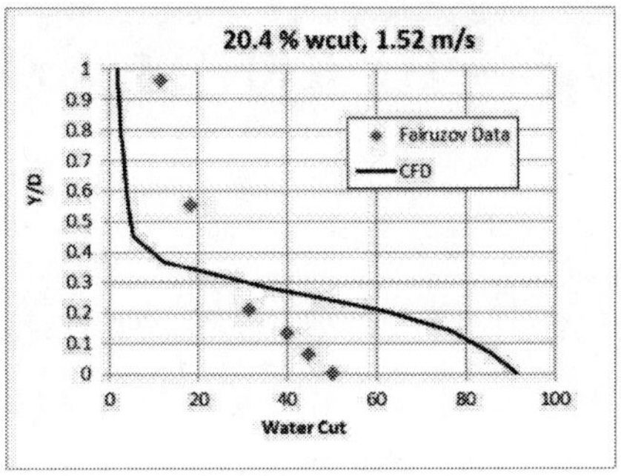

- Excellent match except at bottom when water can form a 100% stratified layer
- At high velocity 1.52 m/s, CFD model shows a much higher water separation at the bottom, indicating the model is more conservative at high water cuts
- However, the slight velocity increase from 1.26 to 1.52 m/s shows a dramatic reduction in oil water separation, which may indicates a major flow regime shift not captured by the model
- However, the data can be an outlier (see next)

DATA OUTLIER?

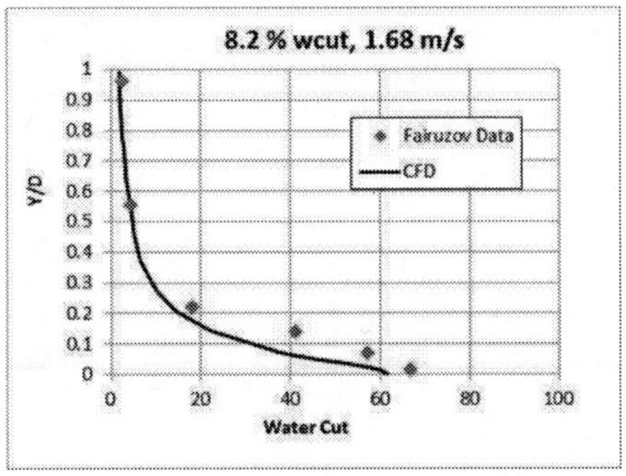

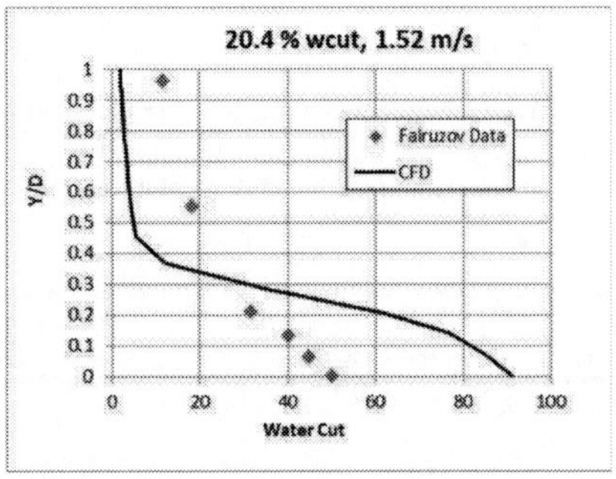

- When comparing the results with a similar velocity and a lower water cut (8.2%), the data shows a **counter-intuitive trend showing a much higher water concentration at the bottom (~70%) is found with a higher velocity and lower water cut.** This seems unlikely and can probably be concluded as a possible experimental error, or extra physics (emulsion?) has been involved that cannot be explained by droplet coalescence alone.

FAIRUZOV EXPERIMENT PREDICTIONS: MULTICORP VS. CFD

Fairuzov conditions: 16″ line

Water cut	Velocity (m/s)	Wettability (Multicorp v4.2)	Wettability (CFD)	Wettability (Exp)	Dmax (mm)	Dmax(mm) CFD	Dcrit (mm)
3.00%	0.91	WW	WW	WW	7.52	3.7	0.9
3.00%	1.13	WW	OW	OW	5.9	1.9	1.33
3.00%	1.89	WW	OW	OW	3.17	1.9	2.87
7.20%	0.61	WW	WW	WW	11.57	>10	0.43
8.20%	1.68	WW	WW	WW	3.71	4.1	2.69
7.20%	1.91	WW	OW	OW	3.22	2.8	2.87
17.70%	1.26	WW	WW	WW	7.88	>10	1.57
20.40%	1.52	WW	WW	OW/WW	7.01	> 10	2.18

- Ohio U. model (Multicorp v. 4.2) is more conservative in predicting water wettability for large ID pipes
- CFD model matches experiments better and is able to distinguish water and oil wettability more accurately
- CFD model shows that the d_{crit} criteria is only an approximation. Water may not fall out even with larger droplets

- Oil
  - Density: 0.851 SG
  - Viscosity: 4.84 cp @100 °F
- Water
  - Density: 1 SG
  - Viscosity: 0.65 cp @ 100 °F
  - Interfacial surface tension: 0.03 n/m
- Pipe ID: 16″
- Pipe length: 284 m

492594 Modeling Pigment Deagglomeration for Industrial Coatings Production

Abstract:

Pigment particle size distribution control is essential in high-value industrial and architectural coatings applications. A common approach to achieve the desired particle size is to disperse a pigment into a slurry with a high speed mixer, typically equipped with a single sawtooth-style impeller. This equipment delivers high specific energy near the impeller to deagglomerate particles and creates bulk flow in the vessel to draw material into the high shear zone. However, dispersion impellers are typically not optimized and may lead to long cycle times. Predicting the particle size distribution of a pigment grind with modeling can lead to more optimal impeller designs that balance power dissipation and circulation in the tank and reduce cycle times.

We use M-Star Simulation LLC's Digital Mixing Tank software to model agglomerated pigment dispersions and compare to experimental lab-scale dispersions. We monitor the evolution particle size by grind gauge and other analytical measurements. Additionally, we discuss the utility of the software to model other particle-particle interactions.

494278 Investigation of Bubble Induced Turbulence Model By Direct Numerical Simulation

Abstract:

Gas-liquid multiphase flow is commonly encountered in bubble columns, stirred tanks and other reactors. The complexity of multiphase turbulent flow increases the difficulty of design and scale-up of the reactors. CFD simulation is an effective tool to analyze and predict multiphase flow in reactors, which is widely used for aided design of industrial reactors. The development of suitable models and numerical methods is hence a significant task.

Gas-liquid flow can be numerically investigated by various methods. For small-scale simulation of detailed turbulence and bubbles, direct numerical simulation (DNS) can be employed. For reactor scale simulation, the Euler-Euler two-fluid model is commonly used, coupled with the Reynolds-averaged Navier-Stokes (RANS) method. The two-fluid model is preferable for industrial multiphase flow simulation due to its simplicity. For gas-liquid flow, the presence of bubbles will affect the structure and intensity of local turbulence. In turn, the turbulence also affects the bubble behavior. Therefore, suitable models should be developed taking the bubble induced turbulence into account, based on the two-fluid model.

There are two ways to introduce the effect of bubbles into liquid turbulence. One way is additional viscosity method, in which an additional viscosity due to bubble motion is added to calculate the effect viscosity [1]. The other way is an additional source term method, in which additional source terms are added to the turbulence kinetic and energy dissipation equations [2]. Both methods are widely used for gas-liquid simulations in stirred tanks and other reactors [3,4]. However, these models include empirical parameters, which are taken on various values in the literature.

In this work, bubble induced turbulence models are investigated by direct numerical simulations. First, gas-liquid flow in a periodic vertical channel is predicted by direct numerical simulation using a front tracking method (Fig.1) [5]. The flow field and other quantities of interest, at every temporal and spatial location, are given by the DNS results. The results are then used to find the various averages and statistical quantities in the model equations. Based on the DNS results and statistical analysis, a bubble induced turbulence model is developed and validated. The model will be used to simulate practice gas-liquid flow in industrial reactors in future works.

Acknowledgements: Financial support from the Li Foundation, the National Natural Science Foundation of China (21427814, 91434126) and National Key Research and Development Program (2016YFB0301701) are gratefully acknowledged.

**Corresponding author. Tel: +86-10-62554558; E-mail: chaoyang@ipe.ac.cn (C. Yang)*

Fig.1 Deformable bubbles and vorticity by DNS

Reference

1. Sato Y., Sadatomi M., Sekoguchi K., Momentum and Heat Transfer in Two-Phase Bubble Flow- = 1 * ROMAN I Theory, Int. J. Multiphase Flow, 1981, 7: 167-177

2. Rzehak R., Krepper E., CFD modeling of bubble-induced turbulence, Int. J. Multiphase Flow, 2013, 55: 138-155

3. Islam A., Adoo N., Bergstrom D., Prediction of Momo-Disperse Gas-Liquid turbulent flow in a vertical pipe, Int. J. Multiphase Flow, 2016, 85: 236-244

4. Wang W., Mao Z.-S., Numerical Simulation of Gas-Liquid Flow in a Stirred Tank with a Rushton Impeller, Chinese J. Chem. Eng., 2002, 10(4): 385-395

5. Tryggvason G., Scardovelli R., Zaleski S., Direct Numerical Simulations of Gas-Liquid Multiphase Flow, Cambridge University Press, 2011

501504 CFD Simulation of a Pulse Jet Mixed Vessel

Abstract:

Pulse Jet Mixed (PJM) vessels have been used to process nuclear waste at Sellafield and are part of the flowsheet for the pretreatment and vitrification of Hanford high level tank waste. PJM vessels contain a series of downward directed pulse tube mixers that periodically discharge fluid into the vessel and are then refilled with vessel fluid, with the cycle being repeated until the process mixing/blending objective is met. During the drive phase, positive air pressure drives the fluid from the narrowed tips of the pulse tubes. These high intensity jets are discharged onto the vessel bottom, travel radially inward and merge to create a central up wash plume that mixes the vessel contents as it dissipates. During suction, negative air pressure pulls the vessel fluid back into the tubes. The air to operate the pulse tubes is supplied from outside leaving the black cell free of moving parts and maintenance. An off normal incident known as "overblow" can occur if a drive to suction phase timing mismatch causes drive air to be discharged into the vessel fluid. Large bubbles travel upward through the vessel into the headspace causing intense fluid agitation. Many of the Hanford vessels will process slurries with fast settling solids while others are relatively dilute. In this study we modeled the turbulent velocity field in a pure fluid during normal PJM operation and in the presence of "overblow" to gain preliminary insight into vessel operations, modeling strategies and design of experiments for model validation.

A Computational Fluid Dynamics (CFD) model was developed for a 1 m diameter pilot scale PJM vessel containing 4 pulse tubes. The vessel fluids were water and a more viscous 30 cP Newtonian liquid of the same density, chosen to bound slurry viscosity. To account for fluid-air interactions, fully transient, 3-D Volume of Fluid (VOF) simulations of the RANS equations were carried out with ANSYS Fluent using the realizable k-e turbulence model with enhanced wall treatment. For normal operation, the pressure profile in the air line feeding the pulse tubes was set to mimic typical cyclic operations. During overblow, pulse tube operation ceased once gas was discharged into the vessel but the evolution of the velocity field was tracked until all bubbles had escaped into the vessel head space. A mesh independence study was conducted and the effect of air compressibility was studied.

The results for normal operation show extremely complex flow patterns throughout the vessel. During drive, mixing layer vortices form between the central up wash plume and the pulse tube bodies and travel upward to the surface. There is weak downward flow close to the vessel walls. During suction, fluid is pulled downward from all vessel locations. There are numerous regions of flow reversal. Increased viscosity impacts the upper regions of the central up wash plume. The mean velocity, deformation and dissipation fields will be discussed in detail. Several overblow scenarios will be presented and contrasted. The effect of air compressibility and bubble expansion on fluid agitation will be shown. Application of the results to Hanford waste treatment plant process operations will be discussed.

510547 Simulation of gas-solid flows in large-scale systems using the direct simulation Monte Carlo method

Abstract:

In this work, the direct simulation Monte Carlo (DSMC) method is developed as a framework for simulating large scale systems with complex physics and hydrodynamic phenomena. While the discrete element method is a Lagrangian approach that can readily incorporate complex particle-scale physics, its use is limited to relatively small systems because of computational costs. Continuum methods such as the two-fluid model can simulate large-scale systems, but there are challenges incorporating particle-scale physics in an Eulerian framework. The DSMC method is a Lagrangian approach that retains the ability to incorporate complex particle-scale physics but is more computationally efficient than discrete element methods. The DSMC framework tracks computational particles that represent many real particles, and the representative particles can experience the same dynamics as real particles. Collisions between computational particles are modeled based on the kinetic theory of granular gases, and the DSMC approach stochastically solves the collision integral of the Enskog equation. Therefore, the DSMC method does not require simplifying assumptions that are often used to derive continuum closures, such as closeness to equilibrium and small Knudsen numbers. To validate the DSMC implementation, we show that the method accurately predicts the onset of the clustering instability in a homogeneous granular gas. The kinetic theory of granular gases breaks down at high solids concentration because of enduring multi-body contacts. To overcome this limitation, the DSMC method is coupled to a deterministic discrete parcel method when the granular flow transitions to the frictional regime. The DSMC approach is then used to model a fluidized bed and industrial scale riser. The simulation space spans the dense to dilute flow regimes and the results are compared to discrete element simulations.

Keywords: Particle technology and fluidization, computational fluid dynamics

499026 Operating Condition Optimization on Mass Transfer in Aerated Stirred Tank Fermentors

Abstract:

Oxygen is a critical component of aerobic fermentations and is sparingly soluble in aqueous solutions. Due to the slow mass transfer of oxygen from the bubbles into the liquid the rate of cell metabolism depends on the rate of oxygen supply from the gas phase. Mass transfer across gas-liquid interfaces plays also a vital role in many other reaction systems. The mass transfer is a function of the gas-liquid surface area which is increased with high agitation rates. The shear rates are increased resulting in bubble brake-up. With the agitation rate the power consumption is increased as well. Micro-organisms, such as yeast, are damageable and thus limit the shear rates, too.

Traditionally dimensionless correlations are used to improve the design of production scale stirred tank fermenters. These aid incorporating many important design variables of the materials and geometry. However, design methodologies available for stirred tank fermenters are insufficient and many times contain errors. By making assumptions on the system correlations often omit important influences on the mass transfer. Using computational fluid dynamics enables us to make fewer assumptions on the geometry and physics resulting in a better insight into the system. This enables the engineer to make meaningful decisions with fewer constraints.

In the current state of industrial engineering a typical vessel is going to be operated yielding various products. Therefore the tank geometry is not designed from scratch but instead has to be used with little alternations. In these or similar scenarios the operating conditions - such as agitation rate or mass flow rates - are a mayor leverage increasing the performance.

In this work the performance of a stirred tank fermenter is going to be optimized using CFD simulations in combination with design explorations. STAR-CCM+ as a finite volume based physics solver and HEEDS exploring the design space are the tools used. After validating the simulation a Pareto front optimization with power consumption vs. mass transfer rate will be presented.

References:

Aglave, R., & Eppinger, T. (2016, December). Influence of Bubble Size Distribution in Gas-Liquid Flows. In *High Performance Computing Workshops*. IEEE.

Doran, P. M. (1995). *Bioprocess engineering principles*. Academic press.

Sorenson, K. L. (2010). *Comparative studies on oxygen mass transfer for the design and development of a single-use fermenteor*. Utah State University.

Davis, R. Z. (2009). *Design and scale-up of production scale stirred tank fermenteors*. Utah State University.

Predicting Temperature Profile of Partially Filled Twin Screw Extruder Using 3D Multiphase CFD Model

Chang Kai (Lance) Wu[1] and Hossam Metwally[2]

[1]Dow Chemical Company
Performance Silicones, Process R&D
Engineering Sciences, Associate Process Engineering Specialist
Midland MI

[2]Ansys Inc.
Principal Engineer
Ann Arbor, MI

dowcorning.com

Outline

- Introduction
- Modeling Challenges
- Mesh Method
- Model Setup
- Results
- Conclusions
- Future Work

2

Dow Chemical Company

9 Large Global R&D Centers
>70 Regional Customer Support Sites

6,000 R&D Employees and a US$1.6 Billion R&D Investment in 2015

DOW CORNING

3

ANSYS is the Simulation Leader

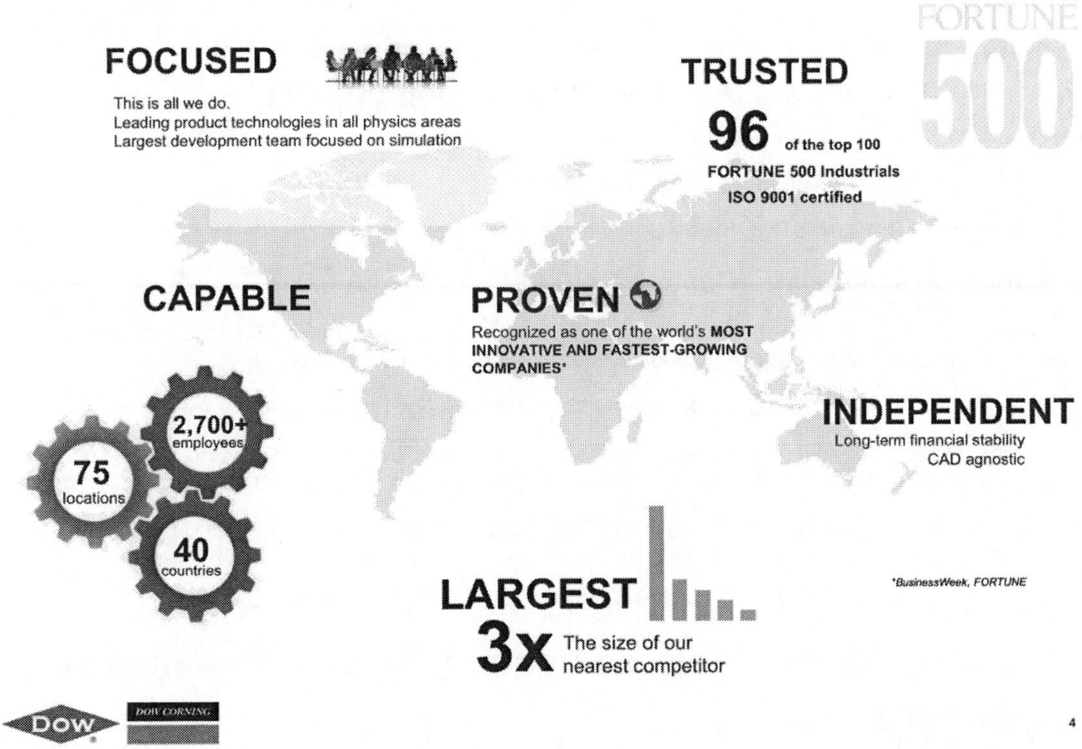

DOW CORNING

4

Twin Screw Extruders

- Continuous modular process, practically all thermoplastics, i.e. PE, PP, PVC, etc
- High viscous polymer, i.e. PTFE, UHMWPE
- Good mixing (Distributive/Dispersive Mixing), Heat Transfer (Heat/Cooling), Filler Incorporation, Reactive Extrusion, Devolatilization, Emulsification…

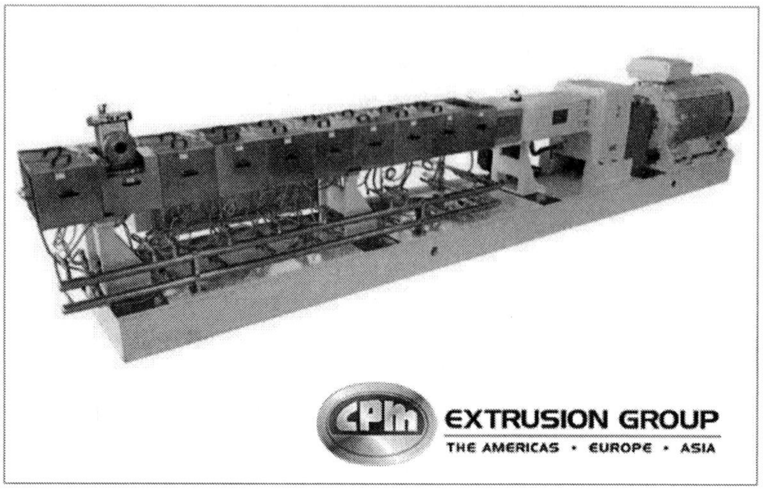

DOW DOW CORNING Photos courtesy of Century Extrusion, 2412 W Aero-Park Ct, Traverse City, MI 49686

5

Modeling Challenges

Modeling Screw Extrusion is difficult because constant moving and deforming fluid volume (meshing problem)

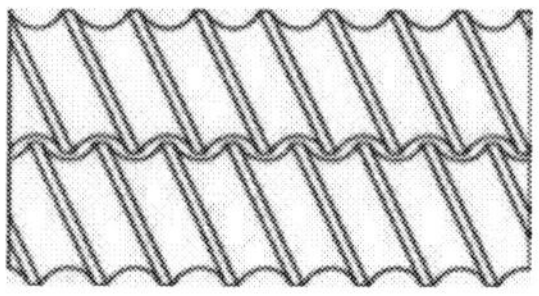

Single-phase modeling (100% full barrel):

- Polyflow, Ludovic, WinTXS, SIGMA,…many more

Multi-phase modeling (Custom Codes):

- *Voorde, J. Vande, and Jan Vierendeels. "A grid manipulation algorithm for ALE calculations in screw compressors." 17th AIAA computational fluid dynamics conference. 2005.*

(differential domain decomposition and grid manipulation)

- *Rane, Sham, and Ahmed Kovacevic. "Algebraic generation of single domain computational grid for twin screw machines. Part I. Implementation." Advances in Engineering Software 107 (2017): 38-50.*

(algebraic, adaptive, deforming grid, SCORG: Screw Compressor Rotor Grid Generation)

Custom codes require deep knowledge

Commercial Available Solution in Fluent (overset mesh)

DOW DOW CORNING

6

Mesh Method - Overset Meshing

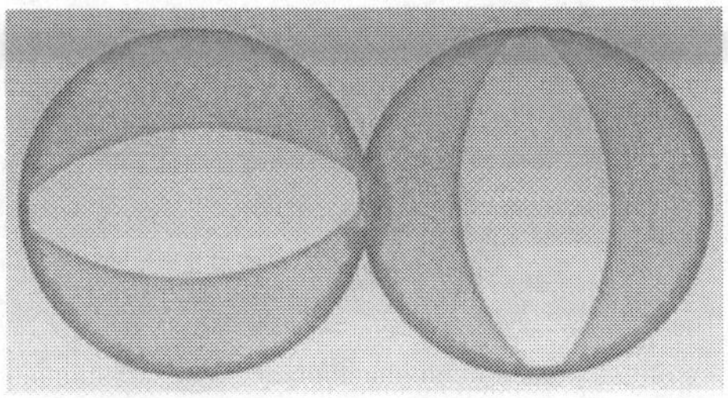

Videos

Overset mesh problems can be set up and solved for (Built in since v17.0**):**

- Steady and transient, 3D and 2D planar
- Pressure Based coupled solver
- Incompressible density method
- Single-phase or VOF multiphase
- Heat transfer
- k-epsilon and SST k-ω turbulence models
- **BETA**: moving mesh, compressible flow, VOF with surface tension, pressure-far-field BC, Workbench support, Pressure-Based segregated algorithms

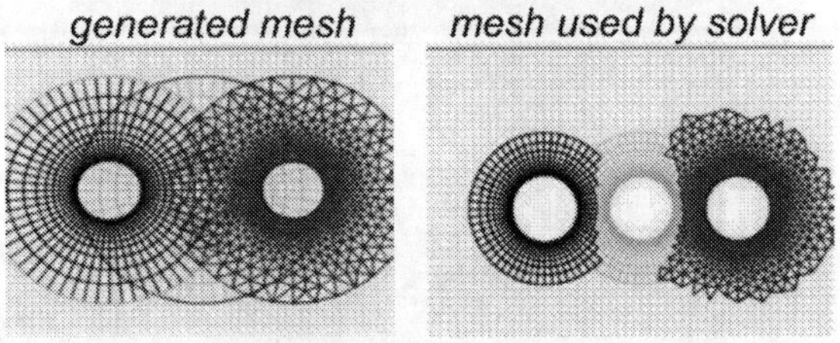

Overset Method: Comparison with Conformal Mesh

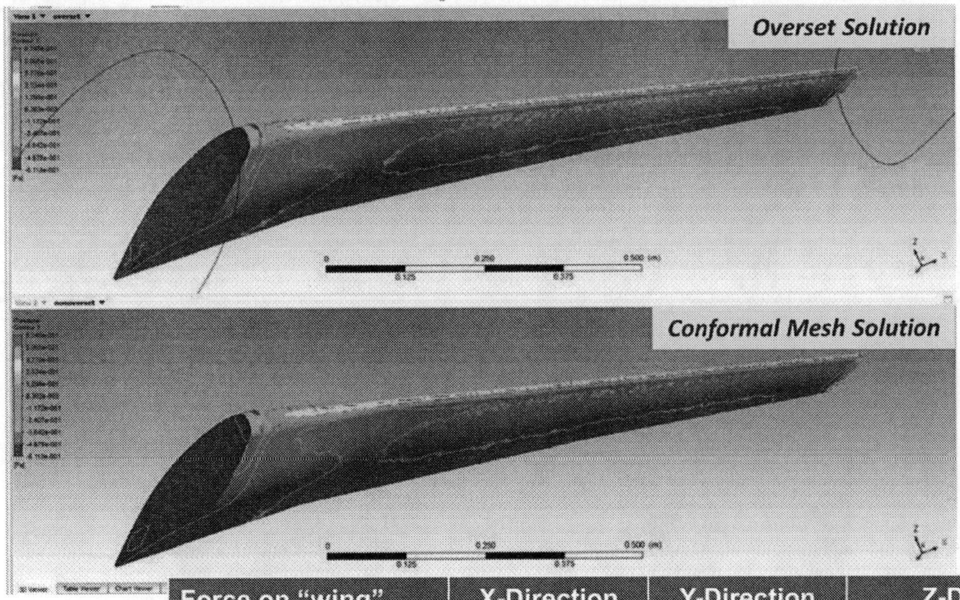

Force on "wing"	X-Direction	Y-Direction	Z-Direction
Conformal Mesh	-0.0164807 N	0.00201411 N	0.0059650 N
Overset Mesh	-0.0165179 N	0.00206875 N	0.0060381 N

Model Setup

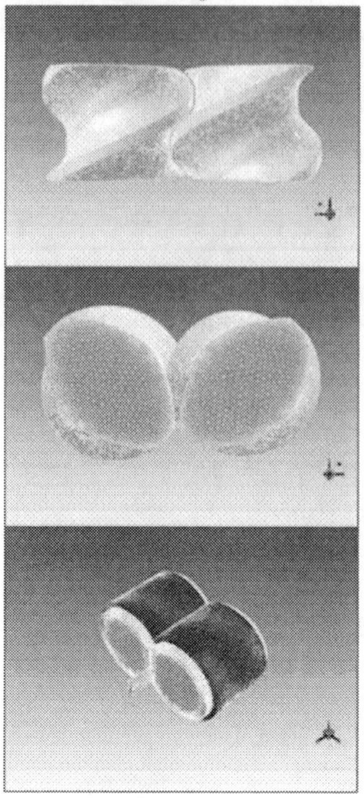

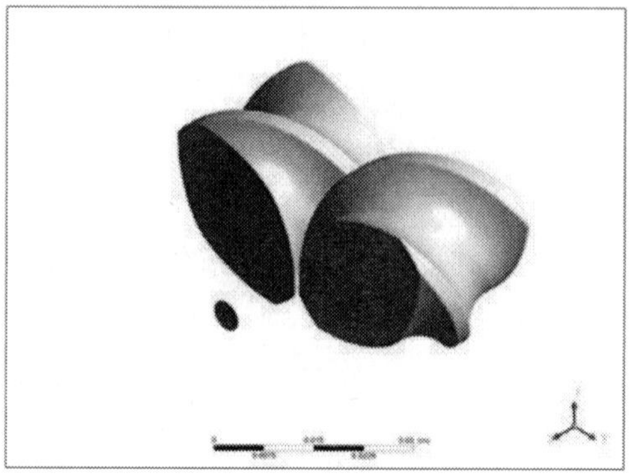

Video

- Fluent 18.2.0
- Standard k-epsilon turbulence model standard wall functions
- Pressure based model with coupled Pressure-Velocity solver
- Screw rotation = 300RPM
- VOF multiphase
- No energy (planed for following models)

Material Properties:

Silicone, density = 1,080 kg/m3, Viscosity= 50 Pa S

Air, density = ideal gas, Viscosity= 0.5 Pa S

9

Element profile 3D CAD courtesy of Century Extrusion, 2412 W Aero-Park Ct, Traverse City, MI 49686

Results: Gas-Liquid Interface

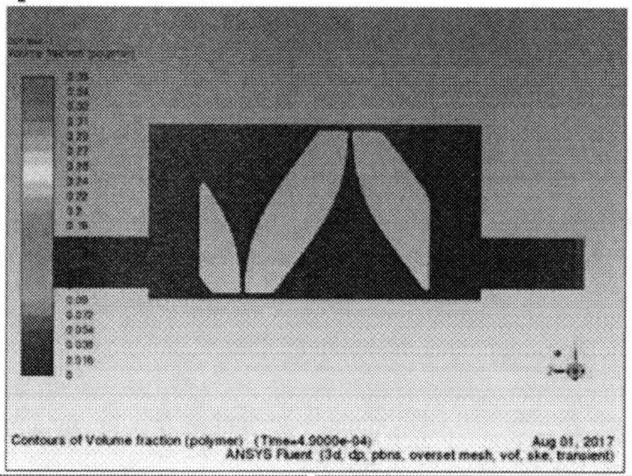

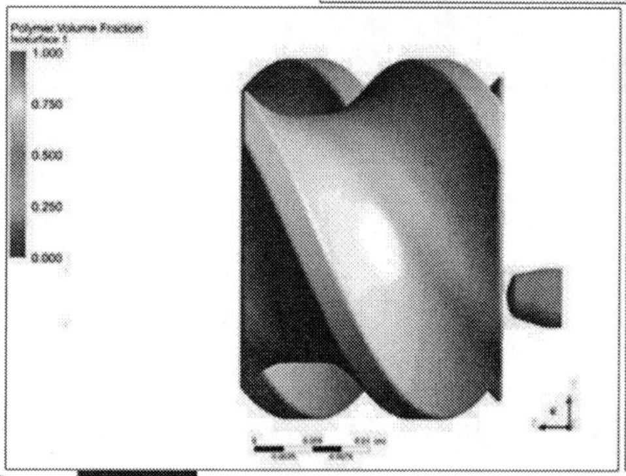

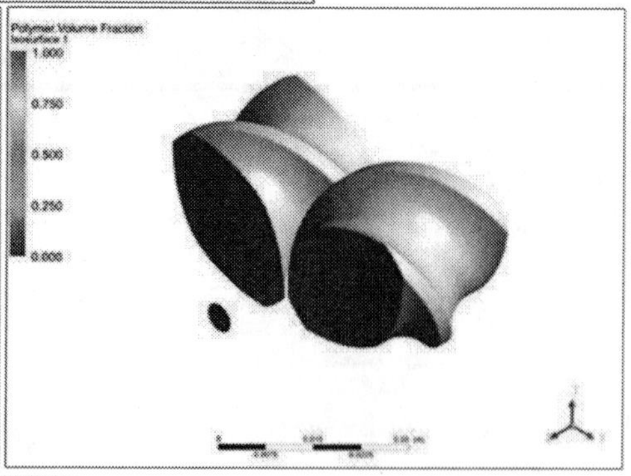

Videos

10

Results: Velocity on Liquid Surface

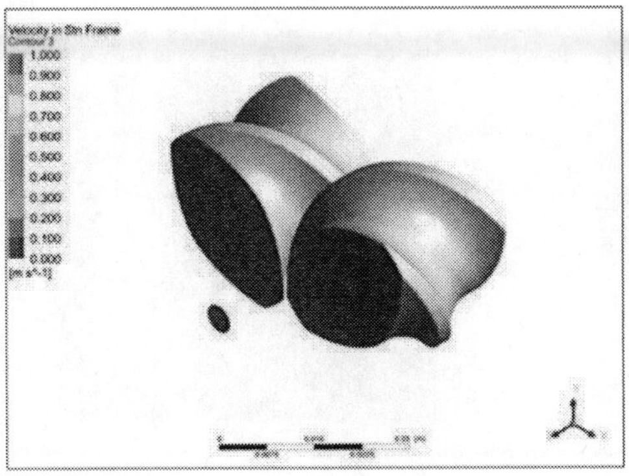

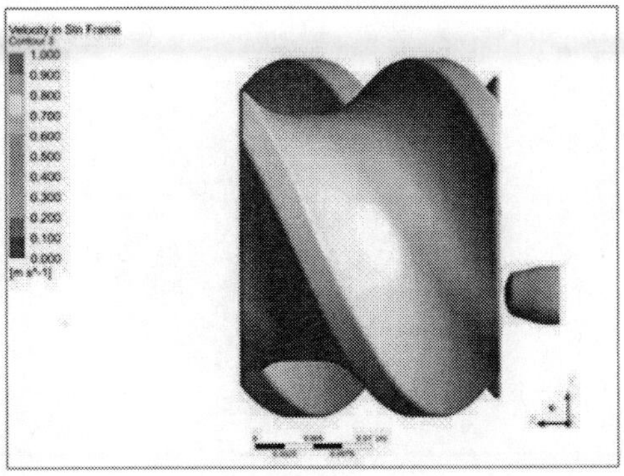

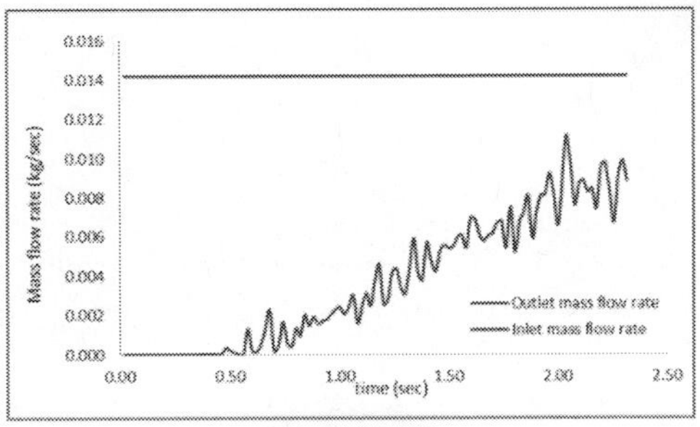

Videos

11

Results: Shear Rate on Barrel Wall Surface

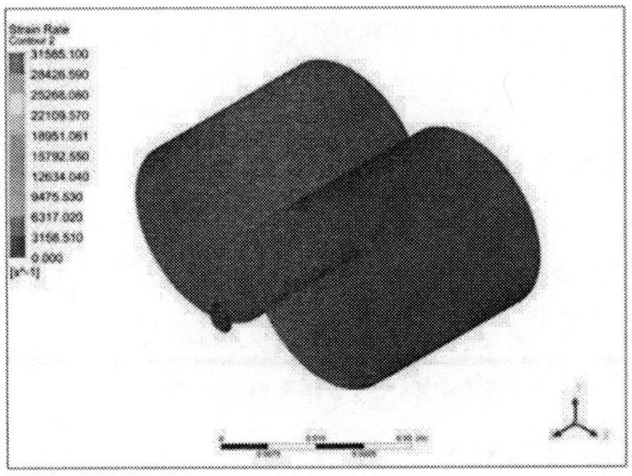

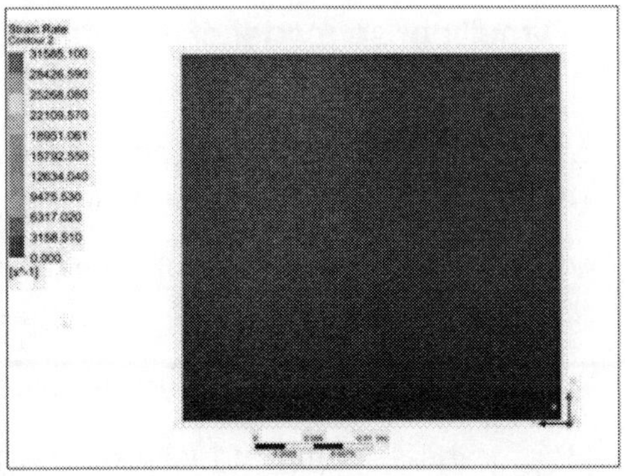

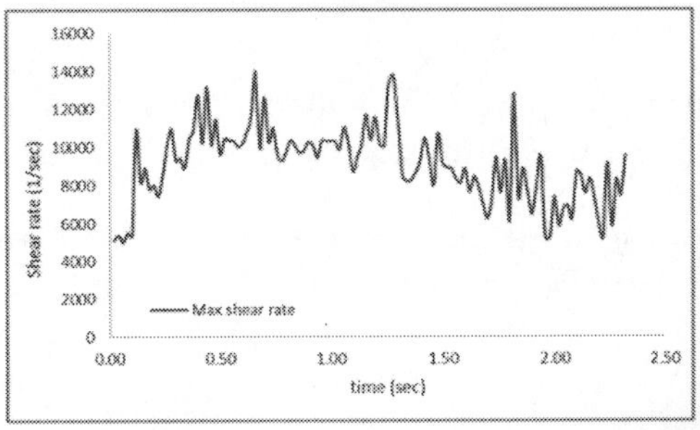

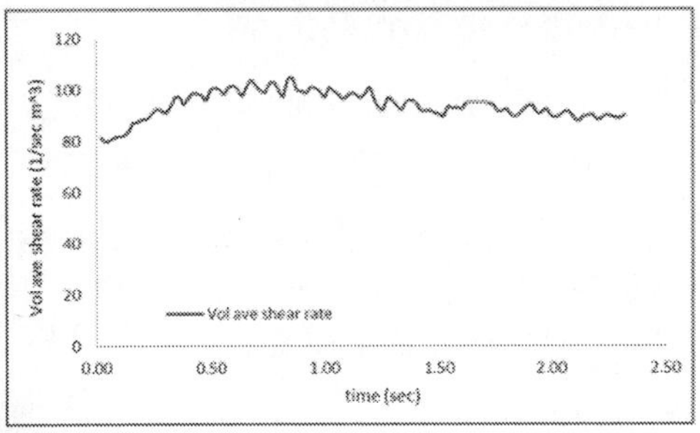

Videos

12

Results: Wetted Area on Barrel Wall Surface

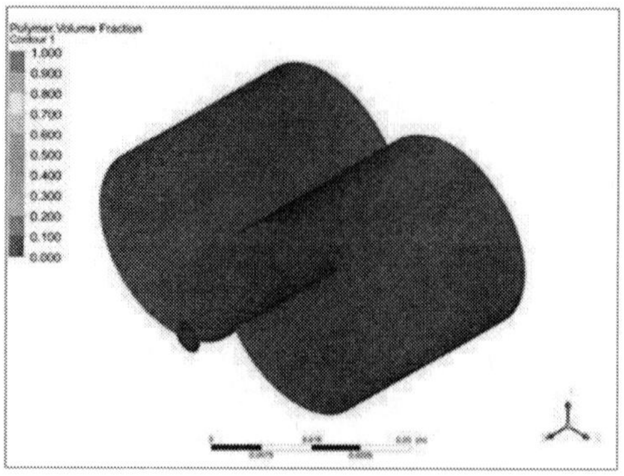

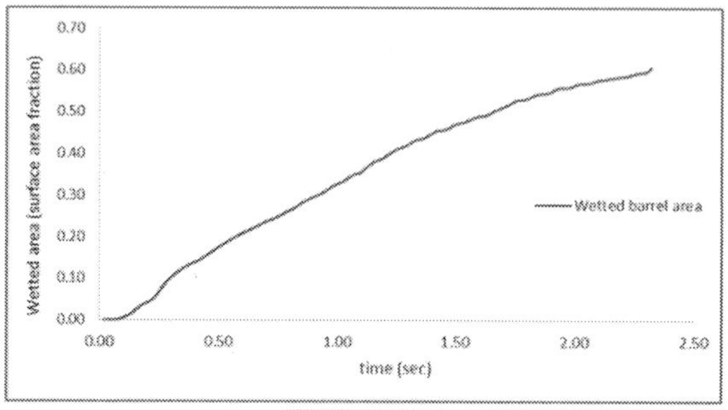

Videos

DOW DOW CORNING

13

Conclusions

- 3D multiphase model of co-rotating fully intermeshing twin screw extruder was created using Ansys Fluent V18.2
- Mesh was created using commercial available overset mesh scheme (built-in Fluent since 17.0 version)
- Volume of fluid (VOF) multiphase model was used to simulate partially filled barrel volume
- Air viscosity is increased to 0.5 Pa S for easier convergence
- Heat transfer was not included in the current model, but it will be included in following simulations
- Results show predicted gas-liquid interface, fluid velocity and shear rate, and the model has not yet reached a steady state after 2 sec of flow time
- Based on the current mass flow rate, model predicted only 60% of barrel surface is wetted, and this value can be used in a simplified conductive heat transfer calculation.

DOW DOW CORNING

14

Future Work

Full scale screw design model with surface heat transfer coefficient prediction

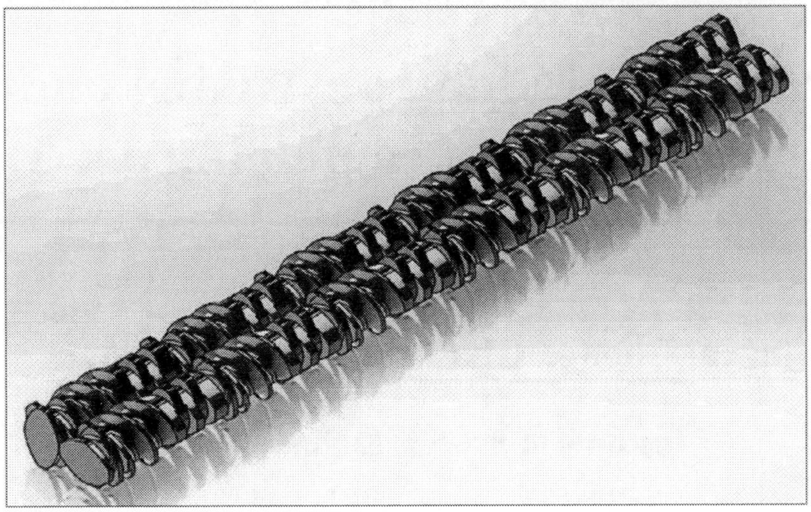

15

Element profile 3D CAD courtesy of Century Extrusion, 2412 W Aero-Park Ct, Traverse City, MI 49686

lance.wu@dowcorning.com
hossam.metwally@ansys.com

Thank You

中国科学院过程工程研究所
Institute of Process Engineering, Chinese Academy Of Sciences

2017 AIChE Annual Meeting, Minneapolis

Numerical Study on Mixing and Crystallization in an Impinging Jet Mixer

Jingcai Cheng, **Chao Yang**, **Zai-Sha Mao**

Institute of Process Engineering,
Chinese Academy of Sciences

Outline

- Introduction
- Mathematical Model
- Simulation Procedure
- Results and discussion
- Conclusions

Introduction

◆ **Impinging jet crystallization:** produce high quality final product with **desired size, narrow and unimodal CSD**, superior crystallinity and purity (Midler et al., 1994; Mahajan & Kirwan, *AIChE J.*, 1996; Liu et al, *AIChE J.*, 2017)

Two Impinging types (Johnson and Prud'homme, *AIChE J.*, 2003)	
➢ **Confined impinging**	**in submerged mode only**
➢ **Free impinging**	**in non-submerged or submerged modes**

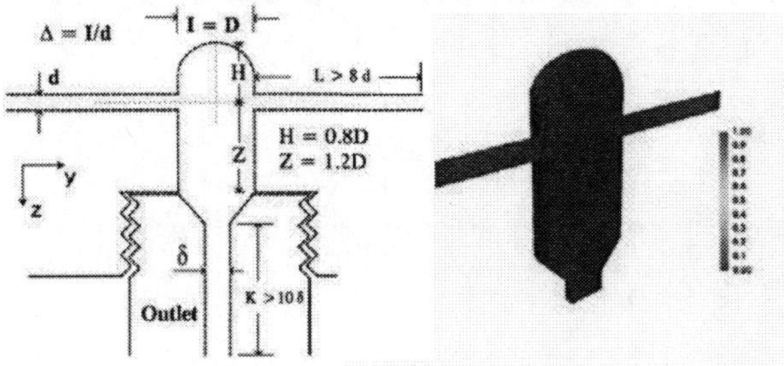

Confined impinging

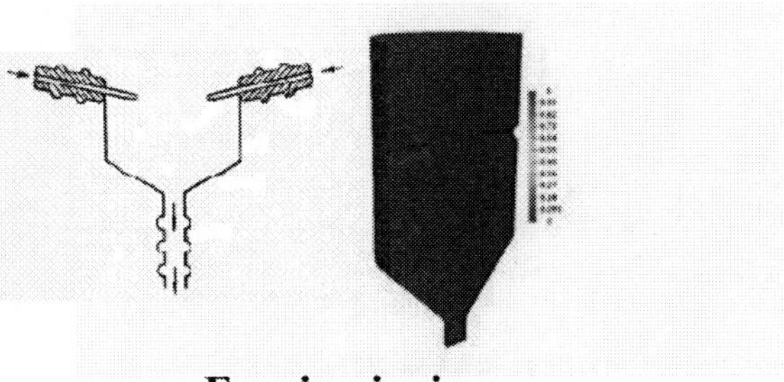

Free impinging

3

Introduction

Why Simulation Tools ?

**: Aid experimental investigation & crystallization processes design

**: Gain insight into processes & reduce development period and cost

CFD-Micromixing-PBE coulped modelling used for **antisolvent crystallization** (widely used in pharmaceutical industry)

◆ **Multiphase CFD:** obtaining **flow fields, turbulent properties** and macromixing characteristics (**mixing** on **reactor scale**)

◆ **Micromixing :** **mixing** on **sub-grid scale** - for most antisolvent crystallization nucleation induction time < mixing time

◆ **PBE:. nucleation, growth/dissolution, aggregation/breakage**

4

Mathematical Model-- Multiphase CFD

- Mixture model: a reduced Eulerian multiphase model

$$\frac{\partial \rho_{\mathrm{m}}}{\partial t} + \nabla \cdot (\rho_{\mathrm{m}} \mathbf{U}_{\mathrm{m}}) = 0$$

$$\frac{\partial (\rho_{\mathrm{m}} \mathbf{U}_{\mathrm{m}})}{\partial t} + \nabla \cdot (\rho_{\mathrm{m}} \mathbf{U}_{\mathrm{m}} \mathbf{U}_{\mathrm{m}}) = -\nabla p + \nabla \cdot \boldsymbol{\tau}_{\mathrm{m}} - \nabla \cdot [\rho_{\mathrm{m}} c_{\mathrm{p}} (1 - c_{\mathrm{p}}) \mathbf{U}_{\mathrm{Sp}} \mathbf{U}_{\mathrm{Sp}}] + \rho_{\mathrm{m}} \mathbf{g}$$

- **Relative (slip) velocity :**

$$\mathbf{U}_{\mathrm{Sp}} = \frac{d_{\mathrm{p}}^2 (\rho_{\mathrm{p}} - \rho_{\mathrm{m}})}{18 f_{\mathrm{drag}} \mu_l} [\mathbf{g} - \frac{D \mathbf{U}_{\mathrm{m}}}{Dt}]$$

- **Drag function:**

$$f_{\mathrm{drag}} = \begin{cases} 1 + 0.15 Re_{\mathrm{p}}^{0.687} & Re_{\mathrm{p}} \le 1000 \\ 0.01825 Re_{\mathrm{p}} & Re_{\mathrm{p}} > 1000 \end{cases}$$

5

Mathematical Model– Micromixing Model

- Three-Environment presumed PDF model (Fox, 2003) :

- **Envt. volume fractions p (p_1, p_2, p_3) & weighted compositions:**

$$\frac{\partial (\alpha_1 \mathbf{p})}{\partial t} + \nabla \cdot (\alpha_1 \mathbf{U}_1 \mathbf{p}) - \nabla \cdot \left(\alpha_1 D_{\mathrm{t}} \nabla \mathbf{p} \right) = \alpha_1 (\mathbf{G}(\mathbf{p}) + \mathbf{G}_{\mathrm{s}}(\mathbf{p}))$$

$$\frac{\partial (\alpha_1 \langle \mathbf{s} \rangle_n)}{\partial t} + \nabla \cdot (\alpha_1 \mathbf{U}_1 \langle \mathbf{s} \rangle_n) - \nabla \cdot \left(\alpha_1 D_{\mathrm{t}} \nabla \langle \mathbf{s} \rangle_n \right) = \alpha_1 (\mathbf{M}''(\mathbf{p}, \langle \mathbf{s} \rangle_1, \ldots, \langle \mathbf{s} \rangle_{N_e}) + \mathbf{M}''_{\mathbf{s}}(\mathbf{p}, \langle \mathbf{s} \rangle_1, \ldots, \langle \mathbf{s} \rangle_{N_e})) + \alpha_1 p_n \mathbf{S}(\langle \boldsymbol{\phi} \rangle_n)$$

- **Other related terms:**

Model variables	$\mathbf{G}$, $\mathbf{M}''$	$\mathbf{G}_{\mathrm{s}}$, $\mathbf{M}''_{\mathrm{s}}$
p_1	$-\gamma p_1 (1 - p_1)$	$\gamma_s p_3$
p_2	$-\gamma p_2 (1 - p_2)$	$\gamma_s p_3$
p_3	$\gamma [p_1 (1 - p_1) + p_2 (1 - p_2)]$	$-2 \gamma_s p_3$
$\langle \mathbf{s} \rangle_3$	$\gamma [p_1 (1 - p_1) \langle \boldsymbol{\phi} \rangle_1 + p_2 (1 - p_2) \langle \boldsymbol{\phi} \rangle_2]$	$-\gamma_s p_3 (\langle \boldsymbol{\phi} \rangle_1 + \langle \boldsymbol{\phi} \rangle_2)$
	$\gamma = \varepsilon_\xi / \left(p_1 (1 - p_1)(1 - \langle \xi \rangle_3)^2 + p_2 (1 - p_2) \langle \xi \rangle_3^2 \right)$	
	$\gamma_s = 2 D_t \nabla \langle \xi \rangle_3 \cdot \nabla \langle \xi \rangle_3 / ((1 - \langle \xi \rangle_3)^2 + \langle \xi \rangle_3^2)$	
	$\langle \xi'^2 \rangle = p_1 (1 - p_1) - 2 p_1 p_3 \langle \xi \rangle_3 + p_3 (1 - p_3) \langle \xi \rangle_3^2$	

6

◆ The general KT high-resolution scheme is validated against analytical solution in homogenous systems first

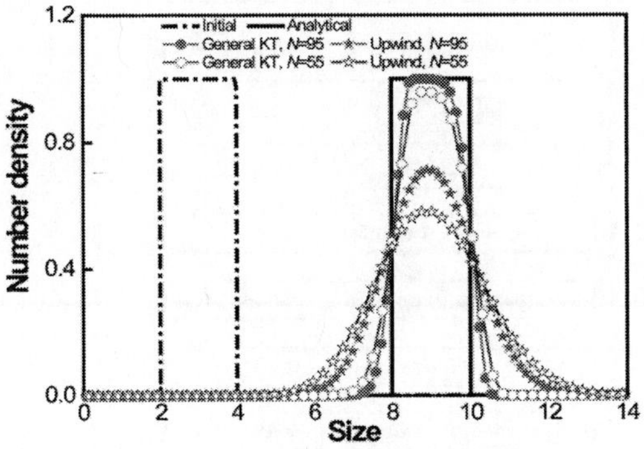

constant growth (uniform size grid)

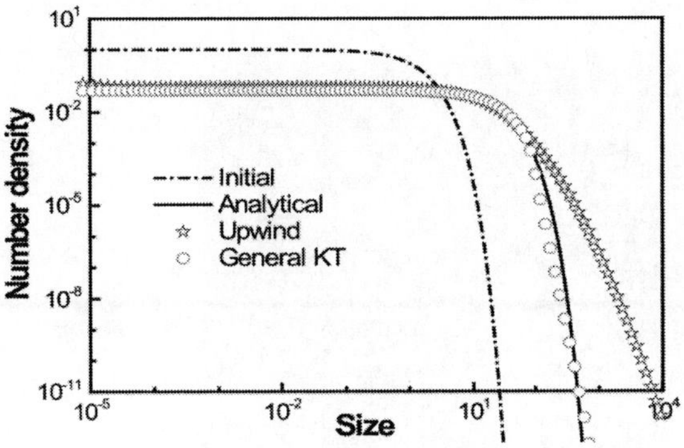

linear growth (geometric size grid):

- Very good agreement with analytical solutions.
- Performing much better than upwind schemes.

Mathematical Model -Discretized PBE Transport Eqns.

◆ Discretized PBE transport eqns. in envt. 3 (on mass basis):

$$\frac{\partial f_{w,\mathrm{j}}}{\partial t}+\nabla\cdot(\mathbf{U}_{\mathrm{p}}f_{w,\mathrm{j}})-\nabla\cdot(D_t\nabla f_{w,\mathrm{j}})=\frac{\rho_c k_v}{4\Delta x_i}\left(x_{j+1/2}^4-x_{j-1/2}^4\right)\left[0^{j-1}B+(-(H_{j+1/2}-H_{j-1/2}))\right],$$

$$j=1,2,\ldots,N$$

➢ **Averaged crystal mass per volume of *j*th size bin:**

$$f_{w,\mathrm{j}}=\rho_c k_v\int_{x_{j-1/2}}^{x_{j+1/2}}x^3 f_j\mathrm{d}x=\frac{\rho_c k_v f_j}{4}\left(x_{j+1/2}^4-x_{j-1/2}^4\right)$$

➢ **Crystal growth:**

$$H_{j+1/2}(t)=\frac{u(f_{j+1/2}^{+}(t))+u(f_{j+1/2}^{-}(t))}{2}-\frac{G_{j+1/2}(t)}{2}(f_{j+1/2}^{+}(t)-f_{j+1/2}^{-}(t))$$

$$H_{j-1/2}(t)=\frac{u(f_{j-1/2}^{+}(t))+u(f_{j-1/2}^{-}(t))}{2}-\frac{G_{j-1/2}(t)}{2}(f_{j-1/2}^{+}(t)-f_{j-1/2}^{-}(t))$$

8

Simulation Platform

The coupled Multiphase CFD-Micromixing-PBE is realized by developing solvers in OpenFOAM (a free source code CFD package written in C++)

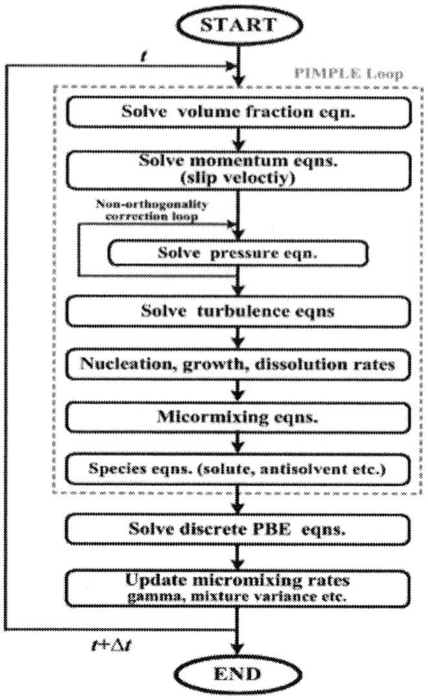

Procedure of the solver in OpenFoam

9

Numerical Procedure

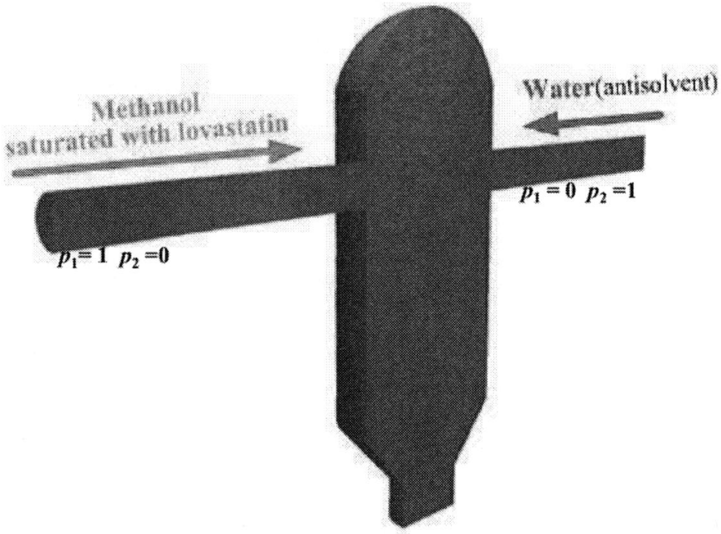

Initial Conditions:

***Envt.* 1**:

pure solution (CH_3OH + lovastatin), $\langle\xi\rangle_1 = 1$

***Envt.* 2**:

pure antisolvent (water), $\langle\xi\rangle_2 = 0$

***Envt.* 3**:

mixing + crystallization

The simulated confined impinging jet mixer

(chamber diameter: **2.38 mm**, inlet tube diameter--**0.5 mm**)

10

Numerical Results –Micromixing Characteristics

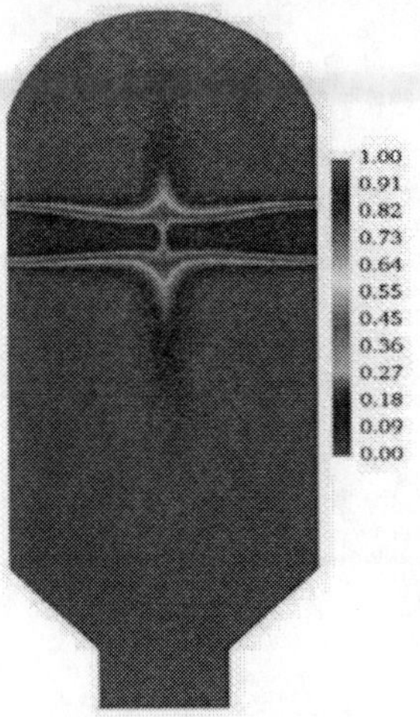

Volume fraction of *Environment* 3 : p_3

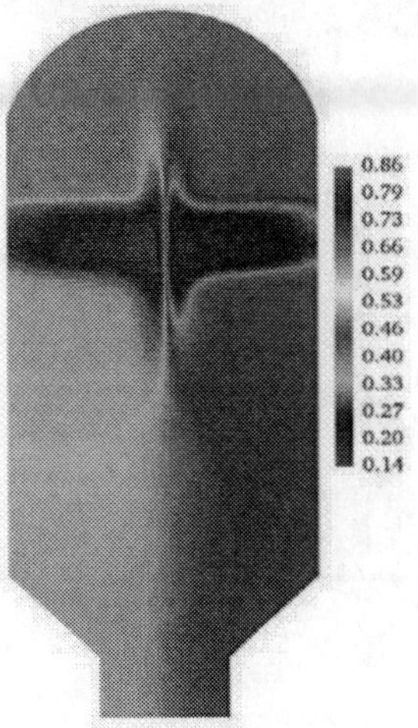

Mixture fraction of *Environment* 3: $\langle\xi\rangle_3$

- **Mixed *Envt.* (p_3) takes over whole domain except impinging region**
- **Amounts of fluids from the two inlets (*Envts.* 1&2) different in the Mixed *Envt.* (p_3)**

11

Numerical Results - Species Concentration

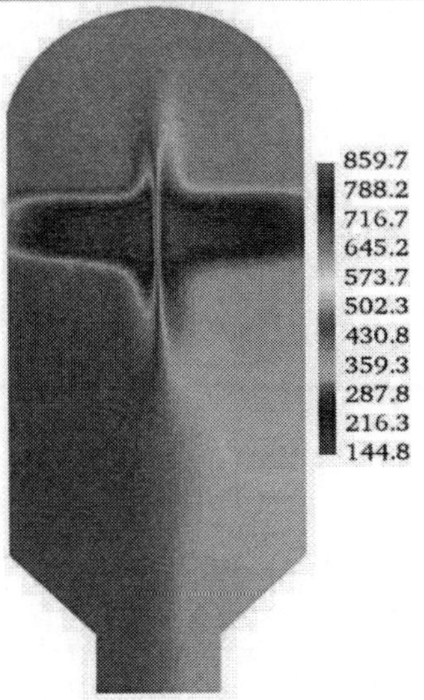

Antisolvent concentration in *Envt.* 3 (kg/m^3)

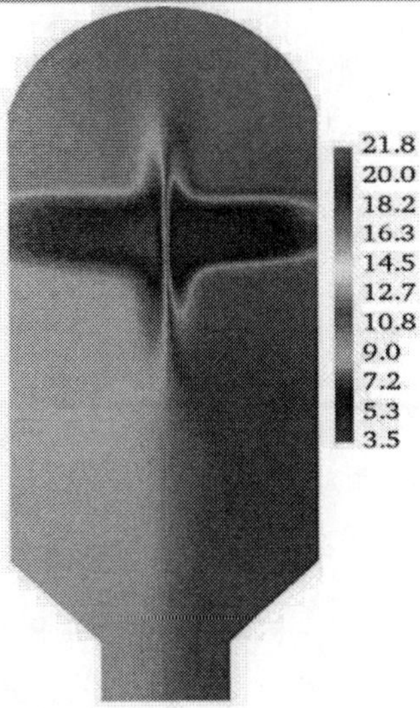

Solute concentration in *Envt.* 3 (kg/m^3)

- **Distributions similar to that for $\langle\xi\rangle_3$ (Mixture fraction of *Envt* . 3)**
- **Less than 50% solute consumed in the exit, the yield needs improving (combining with tube/cooling etc.)**

12

Numerical Results -Crystal Holdup

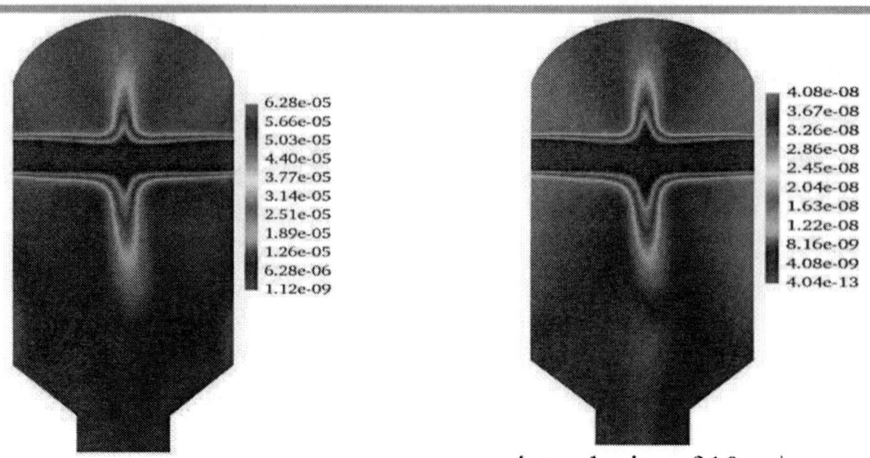

jet velocity of 2 m/s jet velocity of 10 m/s

Distributions of the crystal holdup (vol. fraction) along the symmetry

- **Lower jet velocities have longer residence times for nucleation and growth, thus more large particles;**
- **Higher jet velocities better micromixing, favoring nucleation**

13

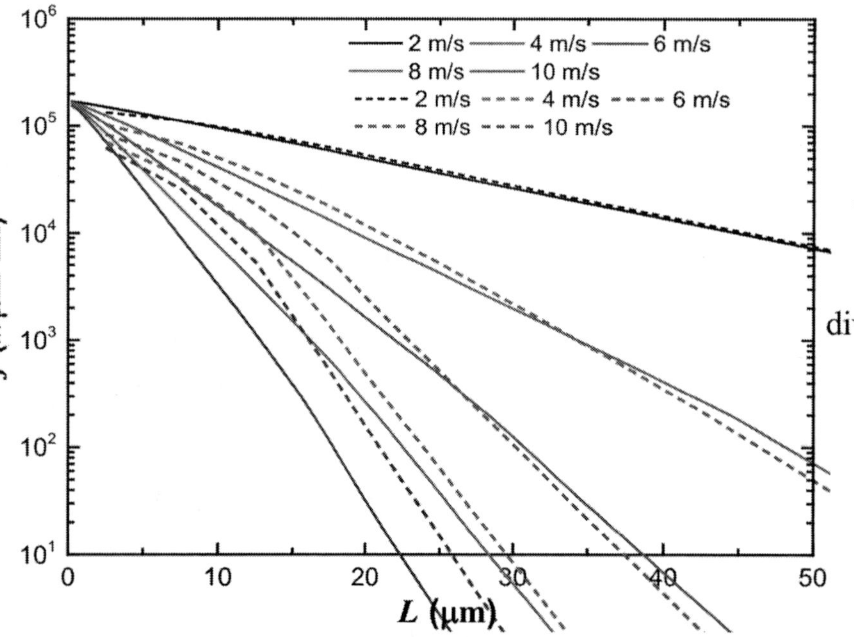

CSDs using uniform size division(dashed lines) and geometric size division(solid lines) at various inlet velocities

- **CSDs from two types of size division** differ considerably at higher jet velocities;
- **Uniform size division** under-predict number density in the smaller size range **(highest number density or steep gradient region exist)**

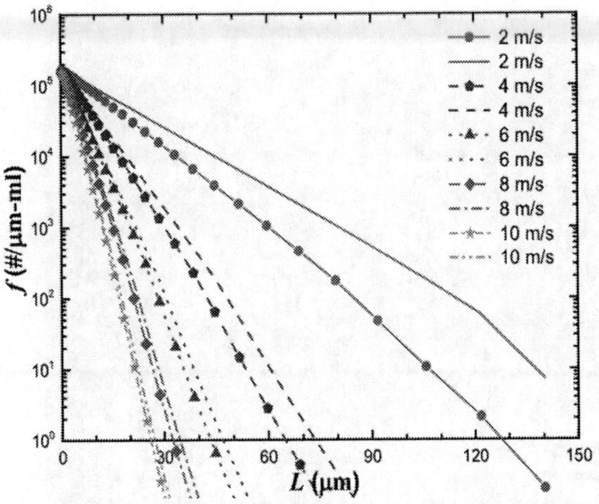

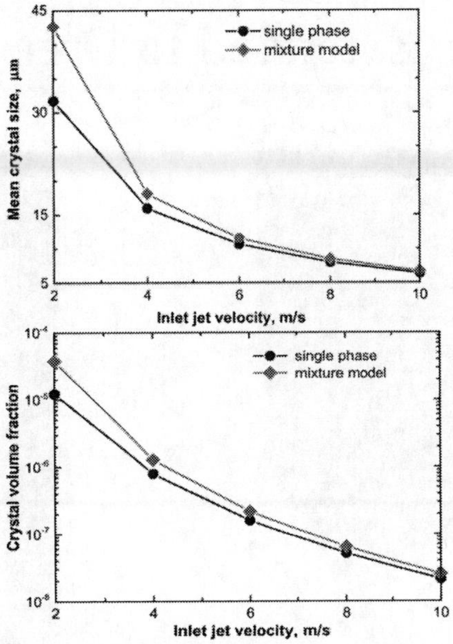

- The existing solid crystals have a **non-negligible effect** on the slurry flow crystallization systems when the mean crystal size **larger than 20 μm**

Numerical Results - Comparison with experiment

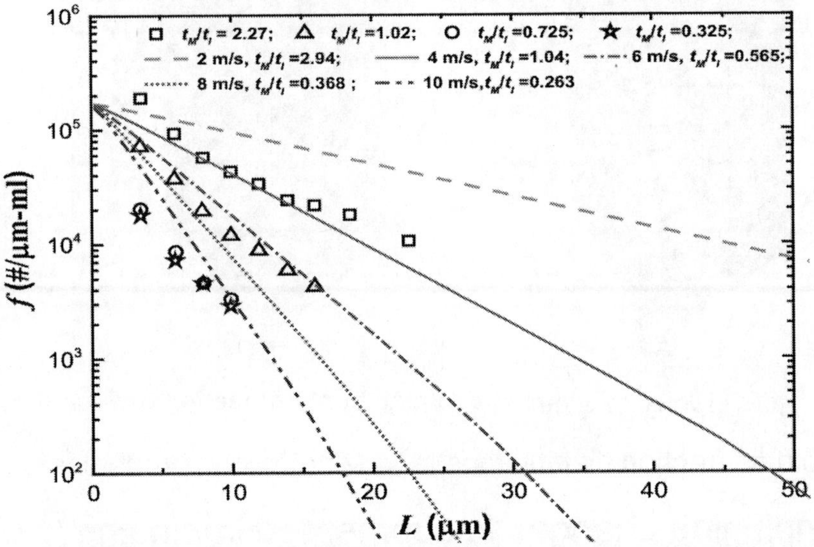

CSDs compared with experiments from (Mahajan & Kirwan, *AIChE J.*, 1996) for similar jet mixer (same cryst. system, but free impinging in non-submerged mode)

- Shape of simulated CSD generally consistent with experimental observations
- CSDs get closer together with increasing jet velocities (**decreasing t_M/t_I**)

16

Extended to Free Impinging Cases

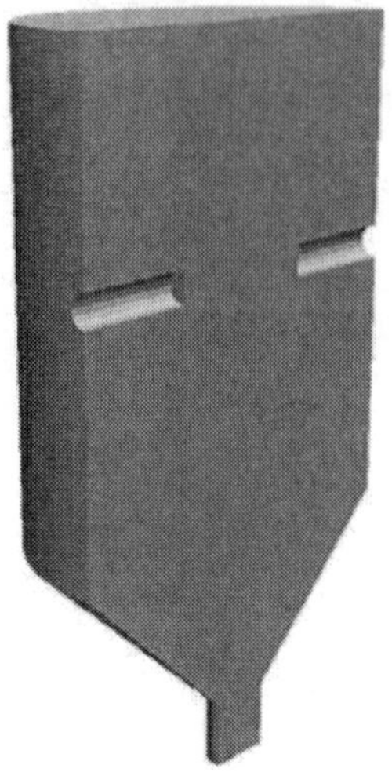

Simulated free impinging jet mixer
(chamber diameter: **25.4 mm**, inlet tube diameter--**3 mm**)

- Volume of Fluid (VOF) to capture the interface between air and solution
- Mixing and Crystallization occur in the dispersed solution phase
- A solver in OpenFOAM also developed

17

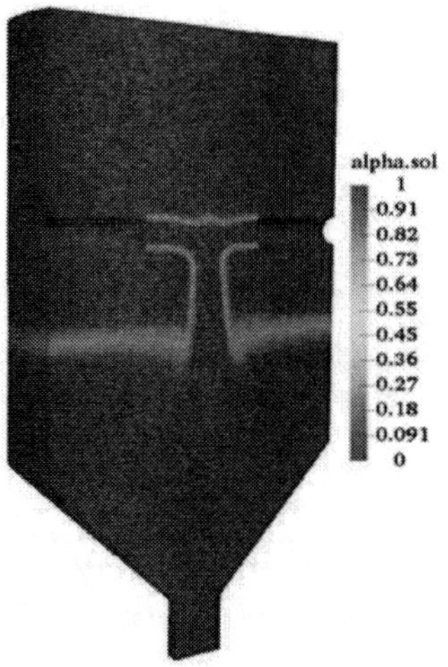

jet velocity of 0.5 m/s

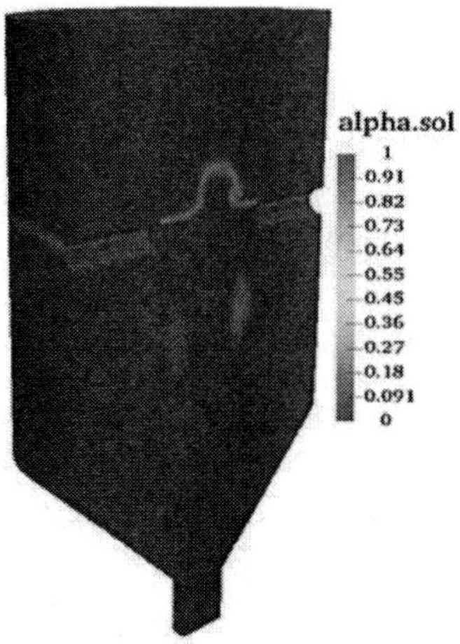

jet velocity of 1 m/s

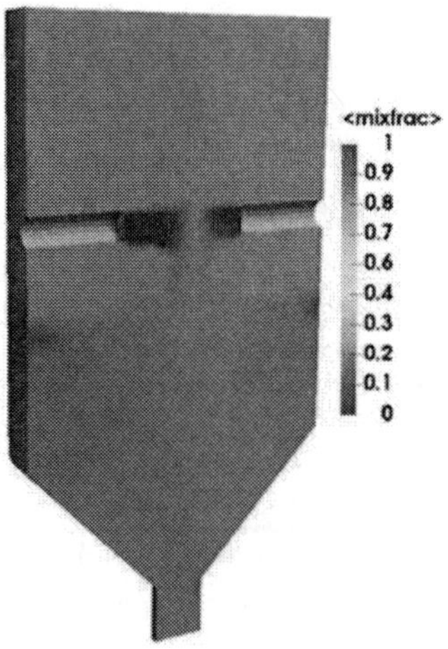

Average mixture fraction (0.5 m/s)

Simulated flow (solution volume fraction distributions) and Micromixing properties

- Non-submerged impinging changes to submerged impinging with increased jet velocities

Work on combining crystallization is ongoing !

18

Conclusions

- **A general multiphase CFD-Micromixing-PBE is developed in OpenFOAM, and simulations conducted successfully on antisolvent crystallization in a confined jet mixer.**
- **Choosing proper size division and considering interaction between crystals and solution improve results considerably in some cases.**
- **Solver in OpenFOAM for free impinging (combining VOF) developed, flow and mixing were numerically studied at present.**

19

Acknowledgements

- **Financial supports from *National Natural Science Foundation* of China and The *Li Foundation Inc.* of USA acknowledged**
- **Advice on this work from Prof. Richard D. Braatz and Dr. Mo Jiang at *Massachusetts Institute of Technology* appreciated**

Thank you for your attention
Questions ?

495396 Power Dissipation in Pharmaceutical Reactors and Vessels Equipped with a Retreat-Blade Impeller

Abstract:

Glass-lined stirred reactors and tanks are commonly used in the pharmaceutical industry because of their corrosion resistance, ease of cleanliness and minimization of product contamination. Most industrial glass-lined tank reactors are provided with a torispherical tank bottom and a retreat curve impeller with low impeller clearance from the vessel bottom. The power, *P*, dissipated by the impeller is a critical process parameter to mixing processes to achieve the desired mixing effect, especially since the power per unit volume, *P/V*, controls directly mass transfer processes and other mixing phenomena. However, little information has been published about the power dissipation and the corresponding power number Po. The objective of this study was to determine experimentally the impeller power dissipation in the vessel and obtain power correlations for a retreat-blade impeller under various types of baffling conditions. In this study the power *P* was measured in fluids of different viscosities and densities at different agitation speeds, and the non-dimensional Power Number, Po, was obtained in a scaled-down version of a typical glass-lined tank reactor for a large range of the Reynolds Number ($1<Re<1,000,000$) for pharmaceutical active ingredient (API) synthesis. Po depends significantly depends on baffling type and Reynolds number, Re. Correlating equations were obtained to predict Po as a function of Re and baffling type. These equations can be used by the industry practitioner to optimize pharmaceutical mixing processes, especially during API synthesis in reactors.

503624 Proven Results of Converting Batch Mixing Systems to Flexible Continuous Inline Blending System

Abstract:

Learn how converting your existing batch mixing operations to continuous inline blending systems can result in

- Reduced capital cost to bring new product variants to market
- Fastest time-to-market for new product launches
- Increased throughput using existing plant equipment
- Rapid changeover with minimum product losses when changing from one product formulation to the next
- Most efficient use of existing plant real estate when introducing new product variants

Learn how to determine if your batch mixing process is a candidate for conversion to an inline blending system by looking at key considerations

- Aqueous / organic solubilities
- Wide differences in viscosity between ingredients
- Large formula turndown and rangeability between products
- Ingredients requiring significant mixing energy
- Staging and handling of flammable ingredients
- Residence time required prior to the introduction of subsequent ingredients
- Cross contamination avoidance

Actual project case studies will be used to illustrate how these results were achieved.

494450 The Effect of Particle Shape in Drum Mixers.

Abstract:

The mixing of particulate materials is critical to the processing of various products produced at industrial scale. The purpose of mixing is to obtain a homogenous product composed of distinct particulate materials, where homogenous refers to a specified mass of each material for a unit volume of the mixture.

Investigating the effect of operational parameters and variations in the properties of input particulate materials on the homogeneity of a product is a tedious and expensive task to perform experimentally. In addition, certain changes may inadvertently not be monitored and their corresponding changes mistakenly attributed to other changes. Here, computational simulation offers a cheap and attractive investigative tool to interrogate changes and consider scenarios that would be impracticable or even impossible to perform experimentally. Early investigative tools relied on Computational Fluid Dynamics (CFD) to provide insight into mixing. However, these simulations lack the resolution to consider changes at particle level on the macroscopic behaviour of the system. Modern approaches include the Discrete Element Method (DEM) that accounts for the particulate nature of materials by explicitly modeling the interaction between individual particles and thus providing insight into the effects of particle properties on the macroscopic mixing. However, the number of particles has been limited and its application to relevant industrial scale simulations hindered by the large computational demand to resolve the particle interactions millions of times per second. As a consequence, the majority of the numerical mixing studies that utilized DEM employed a number of restrictive approximations to reduce the computational burden. These include, the assumption of mono-disperse size distributions, scaling up of particle size and spherical particle shapes.

While recent studies on the GPU have been able to significantly increase the number of particles in a simulation, this has been done at the expense of sacrificing physics by simplifying force models, omitting secondary results such as attrition rates, stress and energy and assuming spherical particle shapes. The omitted secondary information is vital in understanding how operational parameters affect the final product and whether hot spots arise. A major simplification remains the simplification of particle shape as spheres. This assumptions could be overly restrictive for the pharmaceutical industry where feed powders are made from crystalline solids where the shape of the individual particles are distinctly polyhedral with significant angularity. This is significant in that the underlining dynamics of polyhedral particles is vastly different to that of spherical particles, resulting in tighter packing fractions and different segregation rates. The former is critically important in that it is the packing fraction that determines the mass of product in a given volume. In this paper we use the GPU based DEM codes Blaze-DEM and XPS to study the effect of particle shape and particle dispersion in drum mixers using multiple GPUs to simulate millions of polyhedral and spherical particles in a comparative study. We also investigate the use of machine learning algorithms in conjunction with Blaze-DEM to optimize operational parameters for given specifications.

503936 Mixing Award Presentation

Abstract:

This session is an NAMF awards session. It is primarily devoted to a 60 minute presentation by the winner of the Mixing Award given by the North American Mixing Forum. The award is sponsored by Dow Chemical. Additional award presentations by students will also be invited.

AUTHOR INDEX

AUTHOR INDEX